DISCARD

INFERIOR

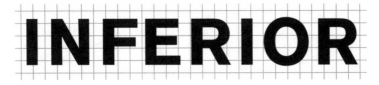

INFERIOR

How **Science Got Women Wrong—**
and the **New Research** *That's*
Rewriting *the* **Story**

Angela Saini

BEACON PRESS
BOSTON

Beacon Press
Boston, Massachusetts
www.beacon.org

Beacon Press books
are published under the auspices of
the Unitarian Universalist Association of Congregations.

21 20 19   8 7 6 5 4

This book is printed on acid-free paper that meets the uncoated paper
ANSI/NISO specifications for permanence as revised in 1992.

Text design and composition by Kim Arney

Library of Congress Cataloging-in-Publication Data

Names: Saini, Angela, author.
Title: Inferior : how science got women wrong and the new research that's
    rewriting the story / Angela Saini.
Description: Boston : Beacon Press, 2017. | Includes bibliographical
    references and index.
Identifiers: LCCN 2016048808 (print) | LCCN 2016057124 (ebook) |
    ISBN 9780807010037 (paperback) | ISBN 9780807071717 (e-book)
Subjects: LCSH: Women's studies. | Women—Psychology. | Women—Physiology. |
    BISAC: SCIENCE / History. | SOCIAL SCIENCE / Gender Studies. |
    SCIENCE / Life Sciences / Evolution.
Classification: LCC HQ1180 .S25 2017 (print) | LCC HQ1180 (ebook) |
    DDC 305.4—dc23
LC record available at https://lccn.loc.gov/2016048808

For my boys,
Mukul and Aneurin

CONTENTS

Introduction I

CHAPTER 1   Woman's Inferiority to Man 13

CHAPTER 2   Females Get Sicker but Males Die Quicker 29

CHAPTER 3   A Difference at Birth .49

CHAPTER 4   The Missing Five Ounces of the Female Brain. 74

CHAPTER 5   Women's Work 96

CHAPTER 6   Choosy, Not Chaste. .120

CHAPTER 7   Why Men Dominate 139

CHAPTER 8   The Old Women Who Wouldn't Die 157

Afterword. 176
Acknowledgments 182
References. 184
Index 201

INTRODUCTION

For centuries, scientists have influenced decision makers on important issues including abortion rights, granting women the vote, and how schools educate us. They have shaped how we think about our minds and bodies and our relationships with each other. And of course, we trust scientists to give us the objective facts. We believe that what science offers us is a story free from prejudice. It is the story of us, starting from the very dawn of evolution.

Yet when it comes to women, so much of this story is wrong.

I was watching a homemade rocket zoom high into the sky. It was a sunny Saturday afternoon and I must have been about sixteen years old, on the playing field of my school in southeast London. Fresh from the nerdy triumph of having been elected chair of the school's first science society, I'd organized a day building small model rockets before shooting them into the air. I couldn't think of anything better. The night before, I desperately calculated whether we even had enough construction materials for the crowds that were sure to come.

I shouldn't have worried. On the day, I was the only one who turned up. My chemistry teacher Mr. Easterbrook, a kind man, stayed and helped anyway.

If you were the geek growing up, you'll recognize how lonely it can be. If you were the female geek, you'll know it's far lonelier. By the time I reached my final years of school, I was the only girl in my chemistry class of eight students. I was the only girl in my mathematics class of about a dozen. And when I decided to study engineering at university, I found myself the only woman in a class of nine.

Things haven't changed much since then. Statistics collected by the Women's Engineering Society in 2016 show that only 9 percent of the engineering workforce in the United Kingdom is female and just over 15 percent of engineering undergraduates are women. Figures from WISE, a campaign in the United Kingdom to promote women in science, engineering, and technology, reveal that in 2015 women made up a little more than 14 percent of their workplaces overall. The picture is similar in the United States: according to the National Science Foundation, although women make up nearly half the scientific workforce, they're underrepresented in engineering, physics, and mathematics.

Standing on that playing field by myself at age sixteen, I couldn't figure it out. I belonged to a household of three sisters, all brilliant at math. Girls stood among boys as the highest achievers at my school. According to the Women's Engineering Society, there's little gender difference in enrollment and achievement in the core science and math subjects at secondary level in UK schools. In fact, girls are now more likely than boys to get the highest grades in these subjects. In the United States, women have earned around half of all undergraduate science and engineering degrees since as far back as the late 1990s.

Yet, as they grow older, fewer women seem to stick with science. At the top, they're in an obvious minority. And this is a pattern that runs as far back as anyone can remember. Between 1901 and 2015, 822 men were awarded a Nobel Prize and only forty-eight women. Of these, sixteen women won the Peace Prize and fourteen won the Prize for Literature. The Fields medal, the world's greatest honor in mathematics, has been won by a woman only once, in 2014 by the Iranian-born mathematician Maryam Mirzakhani.

A couple of years after I graduated from university, in January 2005, the president of Harvard University, economist Lawrence Summers, gave voice to one controversial explanation for this gap. At a private conference he suggested that "the unfortunate truth" behind why there are so few top women scientists at elite universities might in some part have to do with "issues of intrinsic aptitude," that a biological difference exists between women and men. A few academics defended him but, by and large, Summers was met by public outrage. Within a year he announced his resignation as president.

But there have always been gently whispered doubts.

Summers may have dared to say it, but how many people haven't thought the same? That there might be an innate, essential difference between the sexes that sets us apart? That the female brain is fundamentally distinct from the male brain, explaining why we see so few women in the top jobs in science? That hushed uncertainty is what lies at the heart of this book. It's the question mark hanging over us, raising the possibility that women are destined never to achieve parity with men because their bodies and minds simply aren't capable of it.

Even today, we live in the balance of that question, feeding our babies fantasies in pink and blue with the assumption they are deeply different. We buy trucks for our boys and dolls for our girls, and delight when they love them. These early divisions reflect our belief that there's a string of biological differences between the sexes, which perhaps shape us for different roles in society. Our relationships are guided by the notion, fed by many decades of scientific research, that men are more promiscuous and women are monogamous. Even our visions of the past are loaded with these myths. When we picture early humans, we imagine powerful men wandering out into the wilderness to hunt for food, while softer, gentler women stay back, tending fires and caring for children. We go so far as to wonder whether men may be the naturally dominant sex because they're physically bigger and stronger.

Only science has the power to resolve this dark, niggling feeling that never seems to go away no matter how much equality legislation is passed: the feeling that we aren't the same, that, in fact, our biology might explain the sexual inequality that has existed, and continues to exist, across the world.

This is dangerous territory, for obvious reasons. Feminists in particular have passionately argued against the notion that our biology should determine how we live. Many believe that what science says shouldn't make a dent in the battle for basic rights. We deserve an equal playing field, they say, and they're right. But whether or not it sits easily with us, we can't ignore biology either. If biological differences exist, we can't help but want to know. More than that, if we want to build a fairer society, we need to be able to understand these gaps and accommodate them.

The problem is that answers in science aren't everything they seem. When we turn to scientists for resolution, we assume they will be neutral. We think the scientific method can't be biased or loaded against women. But we're wrong. The puzzle of why there are so few women in the sciences is crucial to understanding why, not because it tells us something about what women are capable of but because it explains why science has failed to rid us of the gender stereotypes and dangerous myths we've been laboring under for centuries.

Women are so grossly underrepresented in modern science because, for most of history, they were treated as intellectual inferiors and deliberately excluded from it. It should come as no surprise, then, that this same scientific establishment has also painted a distorted picture of the female sex. This, in turn again, has skewed how science looks and what it says even now.

When I stood on my own on that playing field as a girl, shooting rockets into the air, I was in love with science. I thought it was a world of clear answers, untainted by subjectivity or prejudice. It was a beacon of rationality free from bias. What I didn't yet understand was that I found myself alone because it's not.

If you want to know what science tells us about the female of our species, there's no better place to begin than by understanding the experiences of women working in science today. UNESCO, the United Nations Educational, Scientific and Cultural Organization, which keeps global figures on women in science, estimates that in 2013 just a little more than a quarter of all researchers in the world were women. In North America and Western Europe, female researchers were 32 percent of the population. In Ethiopia, the proportion of female researchers was only 13 percent.

The common trend is for women to be around in high numbers at the undergraduate level but to thin out as they move up the ranks. This is best explained by the perennial problem of child care, which lifts women out of their jobs at precisely the moment their male colleagues are putting in more hours and being promoted. When researchers Mary Ann Mason, Nicholas Wolfinger, and Marc Goulden published a book on this subject in 2013, titled *Do Babies Matter: Gender and Family in the Ivory Tower*, they found that married mothers of young children in the United States were a third less likely to receive tenure-track jobs than married fathers of young children. This isn't a matter of women being less talented. Unmarried, childless women are 4 percent more likely to get these jobs than unmarried, childless men.

The US Bureau of Labor Statistics runs an annual Time Use Survey to pick apart how people spend their hours. Women now make up almost half the labor force, yet in 2014 the bureau found that women spent about half an hour more every day than men doing household work. On an average day, a fifth of men did housework, compared with nearly half of women. In households with children under the age of six, men spent less than half as much time as women taking physical care of these children. At work, on the other hand, men spent fifty-two minutes a day longer on the job than women did.

These discrepancies partly explain why workplaces look the way they do. A man who's able to commit more time to the office or laboratory is naturally more likely to do better in his career than a woman who can't. When decisions are made over who should take maternity or paternity leave, it's also almost always mothers who take time out.

Small individual choices, multiplied over millions of households, can have an enormous impact on how society looks. The Institute for Women's Policy Research in the United States estimates that in 2015 women working full time earned only seventy-nine cents for every dollar that a man earned. In the United Kingdom, the Equal Pay Act was passed in 1970. But today, according to the Office for National Statistics, a gender pay gap of more than 18 percent still exists, although it's falling. In the scientific and technical activities sector this gap is as big as 24 percent.

Housework and motherhood aren't the only things affecting gender balance. There's outright sexism, too. In a study published in 2012, psychologist Corinne Moss-Racusin and a team of researchers at Yale University explored the possibility of gender bias in recruitment by sending out fake job applications for a vacancy of laboratory manager. Every application was identical except that half were given a female name and half a male name. When they were asked to comment on these potential employees, scientists rated women significantly lower in competence and hireability. They were also less willing to mentor them and offered far lower starting salaries. The only difference, of course, was that these applicants appeared to be female.

Interestingly, the authors wrote in their paper, which appeared in the journal *Proceedings of the National Academy of Sciences*, "The gender of the faculty participants did not affect responses, such that female and male faculty were equally likely to exhibit bias against the female student." Gender bias is so steeped in the culture, their results implied, that women were themselves discriminating against other women.

Another study, published in 2016 in the world's largest scientific journal, *PLOS ONE*, looked at how male biology students rated their female counterparts. Cultural anthropologist Dan Grunspan, biologist Sarah Eddy, and their colleagues asked hundreds of undergraduates at the University of Washington what they thought about how well others in their class were performing. "Results reveal that males are more likely than females to be named by peers as being knowledgeable about the course content," they wrote. This didn't reflect reality. Male grades were overestimated—by men— by 0.57 points on a four-point grade scale. Female students didn't show the same gender bias.

The year before, *PLOS ONE* had been forced to apologize after one of its own peer reviewers suggested that two female evolutionary geneticists who had authored a paper should add one or two male coauthors. The paper itself was about gender differences among doctorates. "Perhaps it is not so surprising that on average male doctoral students co-author one more paper than female doctoral students, just as, on average, male doctoral students can probably run a mile a bit faster than female doctoral students," wrote the reviewer.

Another problem in parts of the sciences, the extent of which is only now being laid bare, is sexual harassment. In 2015 virus researcher Michael Katze was banned from entering the laboratory he headed at the University of Washington following a string of serious complaints, which included the sexual harassment of at least two employees. *BuzzFeed News* (which Katze tried to sue to block the release of documents) ran a lengthy account of the subsequent investigation, revealing that he had hired one employee "on the implicit condition that she submit to his sexual demands."

His case wasn't an exception. In 2016 California Institute of Technology suspended a professor of theoretical astrophysics, Christian Ott, for also sexually harassing students. The same year two female students at the University of California, Berkeley, filed a legal complaint against assistant professor Blake Wentworth, who they claimed had sexually harassed them repeatedly, including inappropriate touching. This was not long after a prominent astronomer at the same university, Geoff Marcy, was found guilty of sexually harassing women over many years.

So here, in all the statistics on housework, pregnancy, child care, gender bias, and harassment, we have some explanations for why so few women are at the top in science and engineering. Rather than falling into Lawrence Summers's tantalizing trap of assuming the world looks this way because it's the natural order of things, take a step back. Imbalance in the sciences is at least partly because women face a web of pressures throughout their lives, which men often don't face.

As bleak as the picture is in some places and some fields, the statistics also reveal exceptions. In certain subjects, women tend to outnumber men both at the university level and in the workplace. There are usually more women than men studying the life sciences and psychology. And in some regions, women are much better represented in science overall, showing that culture is also at play. In Bolivia, women account for 63 percent of all scientific researchers. In central Asia they are almost half. In India, where my family originate from (my dad studied engineering there), women make up a third of all students in engineering courses. Iran, similarly, has high proportions of female scientists and engineers. Mathematician Maryam Mirzakhani, the only woman to have won the prestigious Fields medal, was born in Tehran. If women were less capable of doing science than men, we wouldn't see these variations, proving again that the story is more complicated than it appears.

And like all stories, it also helps to go back to the start.

Since its very earliest days, science has treated women as the intellectual inferiors of men. You would see it if you were to travel back to when the major academies of science were first created in Europe, according to Londa Schiebinger, a professor of the history of science at Stanford University and author of *The Mind Has No Sex? Women in the Origins of Modern Science*. In the sixteenth and seventeenth centuries, these academies were founded as forums for scientists, who usually worked independently, to come together and share ideas. Later, they bestowed honors, including membership. These days they also offer governments advice on science policy. Yet these prestigious institutions, so crucial to the growth of modern science, excluded women as a matter of course.

The Royal Society of London, officially founded in 1663 and one of the oldest scientific institutions still around today, failed to elect any women to full membership until 1945. It took until the middle of the twentieth century, too, for the prestigious academies of Paris and Berlin. "For nearly three hundred years, the only permanent female presence at the Royal Society was a skeleton preserved in the society's anatomical collection," notes Schiebinger.

Things got worse before they got better. In its early days, when science was a pastime for enthusiastic amateurs, women had at least some access to it—even if this was only by marrying wealthy scientists and having the chance to work with them in their laboratories. By the end of the nineteenth century, science had transformed into something more serious, with its own set of rules and official bodies. By then, women found themselves almost completely pushed out, says Miami University historian Kimberly Hamlin.

"The sexism of science coincided with the professionalization of science. Women increasingly had less and less access," she explains.

This discrimination didn't just happen higher up in the scientific pecking order. Even assuming she was given the same schooling as a boy, it was unusual for a girl to be allowed into universities or granted degrees until the twentieth century. "From their beginnings European universities were, in principle, closed to women," writes Schiebinger. They were designed to prepare men for careers in theology, law, government, and medicine, which women were barred from entering. Doctors argued that the mental strains of higher education might divert energy away from a woman's reproductive system, harming her fertility.

It was thought that merely having women around might disrupt the serious intellectual work of men, she adds. The celibate male tradition of medieval Christian monasteries continued at the universities of Oxford and Cambridge until late into the nineteenth century. Professors weren't allowed to marry. Cambridge would wait until 1921 to award degrees to women. Similarly, Harvard Medical School refused to admit women until 1945. The first woman applied for a place almost a century earlier.

This doesn't mean that female scientists didn't exist. They did. Many even succeeded against the odds. But they were often treated as outsiders and routinely overlooked for honors. The most famous example is Marie Curie, the first person to win two Nobel Prizes, but nevertheless denied from becoming a member of France's Academy of Sciences in 1911 because she was a woman.

Others are less well known. At the start of the twentieth century, American biologist Nettie Maria Stevens played a crucial part in identifying the chromosomes that determine sex, but her scientific contributions have been largely ignored by history. When mathematician Emmy Noether was put forward for a faculty position at the University of Göttingen during the First World War, one professor complained, "What will our soldiers think when they return to the university and find that they are required to learn at the feet of a woman?" Noether lectured unofficially for the next four years under a male colleague's name and without pay. Albert Einstein described her in the *New York Times* after her death as "the most significant creative mathematical genius thus far produced since the higher education of women began."

Even by the Second World War, when more universities were opening up to female students and faculty, they continued to be treated as second-class citizens. In 1944 the Austrian-born physicist Lise Meitner failed to win a Nobel Prize despite her vital contribution to the discovery of nuclear fission. Her life story is a lesson in persistence. When she was growing up, girls weren't educated beyond the age of fourteen. Meitner was privately tutored so she could pursue her passion for physics. When she finally secured a research position at the University of Berlin, she was given a small basement room and no salary. She wasn't allowed to climb the stairs to the levels where the male scientists worked.

Others, like Meitner, have been denied the recognition they deserve. Rosalind Franklin's enormous part in decoding the structure of DNA was all but ignored when James Watson, Francis Crick, and Maurice Wilkins shared the Nobel after her death in 1962. And as recently as 1974 the Nobel Prize for the discovery of pulsars wasn't given to astrophysicist Jocelyn Bell Burnell, who actually made the breakthrough, but to her male supervisor.

In the history of science, we have to hunt for the women—not because they weren't capable of doing research but because for such a large chunk of time they didn't have the chance. We're still living with the legacy of an establishment that's just beginning to recover from centuries of entrenched exclusion and prejudice.

"I've noted that even the best male minds sometimes become obtuse when they start talking about women—that there is something about gender as a topic that dulls otherwise discerning intellects," writes Mari Ruti, a professor of critical theory at the University of Toronto, in her 2015 book *The Age of Scientific Sexism*.

Sex difference is today one of the hottest topics in scientific research. An article in the *New York Times* in 2013 stated that scientific journals had published thirty thousand articles on sex differences since the turn of the millennium. Be it language, relationships, ways of reasoning, parenting, physical and mental abilities, no stone has gone unturned in the forensic search for gaps. And much of this published work seems to reinforce the myth that the gaps between women and men are huge.

In this book, I unpack some of these studies and interview the people behind them. Some scientists claim that women are on average worse than men at mathematics, spatial reasoning, and anything that requires understanding how systems, such as cars and computers, work. Others say this is because women's brains are structurally different from men's brains. There are also those who insist that men played the dominant part in human evolutionary history because they hunted animals, while women had the apparently less challenging role of staying at home and caring for children. They've argued that males are responsible for humans evolving high intelligence and creativity. Still others say that women experience menopause because men don't find older women attractive.

It can be hard to question their motives. Words that sound deeply objectionable at a dinner party sound remarkably plausible when they're falling from the mouth of someone in a lab coat. But we need to be skeptical. The study you read about in the newspaper telling you that men are better at reading maps than women, for example, may be entirely contradicted by another study on a different population of people, in which women happen to be better map readers. The beautiful brain scan is not the photograph of our thoughts that it sometimes claims to be. And in some branches of science, such as evolutionary psychology, theories can be little more than thin scraps of unreliable evidence strung into a narrative.

If studies seem sexist, occasionally it's because they are. But then, it's impossible not to expect that the very bias that kept women out of science for centuries might have affected the very blood and bones of their work—that it might have prejudiced science's objectivity.

But there's more to this story.

Having more women in science is already changing how science is done. Questions are being asked that were never asked before. Assumptions are being challenged. Old ideas are giving way to new ones. The distorted, often negative picture that research has painted of women in the past has been powerfully challenged in recent decades by other researchers—many of whom are women. And this alternative portrait shows humans in a completely different light.

Today, hidden among the barrage of questionable research on sex differences, we have a radically new way of thinking about women's minds, bodies, and their role in evolutionary history. Fresh theories on sex difference, for example, suggest that the small gaps that have been found between the brains of women and men are statistical anomalies caused by the fact that we are all unique. Decades of rigorous testing of girls and boys confirm that there are few psychological differences between the sexes, and that the differences seen are heavily shaped by culture, not biology. Research into our evolutionary past shows that sexual division of labor and male domination are not biologically hardwired into human society, as some have claimed, but that we were once an egalitarian species. Even the age-old myth about women being less promiscuous than men is being overturned.

This is well-evidenced, careful work that challenges old ideas about what it really means to be a woman. The picture they paint isn't of someone who's weak or subservient. She's not less able to excel in science, nor is she any of the many other softly patronizing adjectives that have been used to mark her apart from men as the more empathic, gentler, fairer sex. This woman is as strong, strategic, and smart as anyone else.

This compelling body of work, rather than pulling women and men farther apart in the gender wars, affirms the importance of sexual equality. It draws us closer together.

———

When I was promoting my first book, *Geek Nation*, I went to the city of Sheffield to give a talk. When I finished, a short, middle-aged man came over to ask some questions in private.

"Where are all the women scientists? Where are the women Nobel Prize winners?" he asked, sneering. "Women just aren't as good at science as men are. They've been shown to be less intelligent." He walked up so close to my face that I was literally backed into a corner. What was a sexist rant quickly became racist, too. I tried to argue back. I listed the accomplished female scientists I knew. I hastily marshaled a few statistics about school-age girls being better at mathematics. But in the end, I gave up. There was nothing I could say for him to think of me as his equal.

How many of us haven't known someone like this? The patronizing boss, the chauvinistic boyfriend, the social media troll, the stranger who thinks a woman's place is in the kitchen? What I wish I had was a set of scientific arguments in my armory to show them that they are wrong. To reinforce that equality isn't just a political ideal but every woman's natural, biological right.

For everyone who has faced the same situation, the same angry confrontation with a person who tells you that women are inferior to men, the same desperate attempt to not lose control but have at hand some real facts and a history to explain them, here they are. In this book I travel through the life stages of a woman, from birth through working life to menopause, to interrogate what science really tells us and the controversies around what remains uncertain.

Despite my personal experience, I didn't set out to write this book with an axe to grind. As a journalist, I have a commitment to the facts. And as someone with an academic background in science and engineering, I wanted to better understand the research. The research I examine spans neuroscience, psychology, medicine, anthropology, and evolutionary biology. Starting in the nineteenth century and running all the way to today, I've tried to find out why so much of what we think of as true is actually unreliable. I investigate the studies that have hit the headlines, claiming to show us that harmful stereotypes about women are backed by science. And at the same time I explore the beautiful, empowering new portrait of women that looks so different from the old one.

This doesn't always make for comfortable reading. The facts are often grayer than people might want them to be. This is simply an account of the science and its controversies as they stand now, chronicling the bitter scientific struggle for the heart and soul of women.

For me, this struggle represents the final frontier for feminism. It has the potential to knock down the greatest barrier that stands between women and full equality—the one in our minds. As anthropologist Kristen Hawkes at the University of Utah put it to me when I interviewed her about her work on menopause for the final chapter of this book, "If you're really paying attention to biology, how can you not be a feminist? If you're a serious feminist and want to understand what the underpinnings of these things are, and where they come from, then *biology*—more science, not less science."

CHAPTER 1

Woman's Inferiority to Man

To prove women's inferiority, antifeminists began to draw not only, as before, on religion, philosophy and theology, but also on science: biology, experimental psychology and so forth.

—Simone de Beauvoir, *The Second Sex*, 1949

The University of Cambridge at the end of summer with the leaves going dry is as beautiful as it must have been when the great evolutionary biologist Charles Darwin was an undergraduate here in the early nineteenth century. Up in the quiet and high northwest corner of the university's library, traces of him still exist. On a leather-topped table in the manuscripts room, I'm holding three letters, all yellowing, the ink faded and the creases brown. Together, they tell a story of how women were viewed in one of the most crucial moments of modern scientific history, when the foundations of biology were mapped out.

The first letter, addressed to Darwin, is written in an impeccably neat script on a small sheet of thick cream paper. It's dated December 1881 and it's from a Mrs. Caroline Kennard, who lives in Brookline, a wealthy town outside Boston. Kennard was prominent in her local women's movement, pushing to raise the status of women (once making a case for police departments to hire female agents). She also had an interest in science. In her note to Darwin, she had one simple request. It was based on a shocking encounter she'd had at a meeting of women in Boston. Someone had taken the position of arguing that "the inferiority of women; past, present and future" was "based upon scientific principles," Kennard writes. The authority that allowed this person to make such an outrageous statement apparently came from no less than one of Darwin's own books.

By the time Kennard's letter arrived, Darwin was only a few months away from death. He had long ago published his most important works, *On the Origin of Species* in 1859 and *The Descent of Man*, which came out twelve years later. They laid out how modern-day humans could have evolved slowly from simpler forms of life by developing characteristics that made it easier to survive and have more children. This was the bedrock of his theories of evolution based on natural selection and sexual selection. And they blasted through Victorian society like dynamite, transforming how people thought about human history. His legacy was assured.

In her letter Kennard naturally assumes that a genius like Darwin couldn't possibly believe that women are naturally inferior to men. Surely his work had been misinterpreted? "If a mistake has been made, the great weight of your opinion and authority should be righted," she entreats.

"The question to which you refer is a very difficult one," Darwin replies the following month from his home in Downe, in Kent. His letter is in a scrawling hand that's so difficult to read that someone had to copy the entire thing word for word onto another sheet of paper, kept alongside the original in the Cambridge University archives. But the handwriting isn't the most objectionable thing about his letter. It's what Darwin actually writes. If polite Mrs. Kennard was expecting the great scientist to reassure her that women aren't really inferior to men, she was about to be disappointed. "I certainly think that women though generally superior to men [in] moral qualities are inferior intellectually," he tells her, "and there seems to me to be a great difficulty from the laws of inheritance, (if I understand these laws rightly) in their becoming the intellectual equals of man."

It doesn't end there. For women to overcome this biological inequality, he adds, they would have to become breadwinners like men. And this wouldn't be a good idea because it might damage young children and the happiness of households. Darwin is telling Kennard that women aren't just intellectually inferior to men, but they're better off not aspiring to a life beyond their homes. It's a rejection of everything Kennard and the women's movement at the time were fighting for.

Darwin's personal correspondence echoes what's expressed quite plainly in his published work. In *The Descent of Man* he argues that males gained the advantage over females across thousands of years of evolution because of the pressure they were under to improve in order to win mates. Male peacocks, for instance, evolved bright, fancy plumage to attract sober-looking peahens. And male lions evolved their glorious manes. In evolutionary terms, he implies, females can happily reproduce no matter how dull they are because they're the ones that give birth. They have the luxury of sitting back and choosing a mate, while males have to work hard to impress them and compete with other males for their attention. In this vigorous competition for women over millennia, the logic goes, men have had to be warriors and thinkers. And this has honed them into finer physical specimens with sharper minds. Women are literally *less evolved* than men.

"The chief distinction in the intellectual powers of the two sexes is shewn by man attaining to a higher eminence, in whatever he takes up, than woman can attain—whether requiring deep thought, reason, or imagination, or merely the use of the senses and hands," he explains in *The Descent of Man*. For Darwin, the evidence appeared to be all around him. Leading writers, artists, and scientists were almost all men. He assumed this inequality reflected a biological fact. Thus, his argument goes, "man has ultimately become superior to woman."

This all makes for astonishing reading now. Darwin writes that if women had somehow managed to develop some of the same remarkable qualities as men, it may have been because they were dragged along on men's coattails by the fact that children happen to inherit a bit of everything from both parents in the womb. Girls, by this process, manage to steal some of the superior qualities of their fathers. "It is, indeed, fortunate that the law of the equal transmission of characters to both sexes has commonly prevailed throughout the whole class of mammals; otherwise it is probable that man would have become as superior in mental endowment to woman, as the peacock is in ornamental plumage to the peahen." It's only a stroke of biological luck, he implies, that has stopped women from being even more inferior to men than they already are. Trying to catch up is a losing game—nothing less than a fight against nature.

To be fair to Darwin, he was a man of his time. His traditional views on a woman's place in society don't run through just his own scientific works but also those of many other prominent biologists of the age. His ideas may have been revolutionary, but his attitudes to women were solidly Victorian.

We can guess how Caroline Kennard must have felt about Darwin's comments from the fiery, long response she sent back. Her second letter is not nearly as neat as her first. She argues that, far from being housebound, women contribute just as much to society as men do. It was, after all, only in wealthier middle-class circles that women tended not to work. For many Victorians, women's incomes were vital to keeping families afloat. The difference between men and women wasn't the *amount* of work they did, but the *kind* of work they were allowed to do. In the nineteenth century, women were barred from most professions as well as politics and higher education.

As a result, when women worked, it was generally in lower-paid jobs such as domestic labor, laundry, the textile industries, and factory work. "Which of the partners in a family is the breadwinner," Kennard writes, "when the husband works a certain number of hours in the week and brings home a pittance of his earnings . . . to his wife; who early and late with no end of self sacrifice in scrimping for her loved ones, toils to make each penny."

She ends on a furious note. "Let the 'environment' of women be similar to that of men and with his opportunities, before she be fairly judged, intellectually his inferior, please."

———

I don't know what Darwin made of Kennard's reply. There's no more correspondence between them in the library's archives.

What we do know is that she was right—his scientific ideas mirrored how society felt at the time, and this was coloring his judgment of what women were capable of doing. Darwin's attitude belonged to a train of scientific thinking that stretched back at least as far as the Enlightenment, when the spread of reason and rationalism through Europe changed the way people thought about the human mind and body. "Science was privileged as the knower of nature," Londa Schiebinger, historian at Stanford University, explains to me. "For women, that nature was described as the characteristics that belong to the private sphere of the home. The nurturing mothers and their job in the state was to educate new citizens, presumably males." Men were portrayed as belonging to the public sphere, where science also happened to live.

By the middle of the nineteenth century, when Darwin was carrying out his research, the image of the weaker, intellectually simpler woman had hardened into a widespread assumption. Society expected wives to be virtuous, passive, and submissive to their husbands. It was an ideal illustrated in a popular verse of the time, "The Angel in the House," by the English poet Coventry Patmore. "Man must be pleased; but him to please / is woman's pleasure," he wrote. Many thought women were naturally unsuited to careers in the professions. They didn't need to have public lives. They didn't need the vote.

When these prejudices met evolutionary biology, they turned out to be a particularly toxic mix, one that would poison scientific research for many decades. Prominent scientists made no secret that they thought women were the inferior half of humanity, the same way Darwin had.

Indeed, it's hard today to read some of the things that famous Victorian thinkers wrote about women and not be shocked. In an article published in *Popular Science Monthly* in 1887, the evolutionary biologist George John Romanes, a friend of Darwin's, patronizes women with his praise of their "noble" and "lovable" qualities, including "beauty, tact, gayety, devotion, wit." He also insists, like Darwin had, that women can never hope to reach the same intellectual heights as men, however hard they try. "From her abiding sense of weakness and consequent dependence, there also arises in woman that deeply-rooted desire to please the opposite sex which, beginning in the terror of a slave, has ended in the devotion of a wife," he writes.

Meanwhile, in the popular 1889 book *The Evolution of Sex*, Scottish biologist Patrick Geddes and naturalist John Arthur Thomson argue that women and men are as different from each other as passive eggs and energetic sperm. "The differences may be exaggerated or lessened, but to obliterate them it would be necessary to have all the evolution over again on a new basis. What was decided among the prehistoric Protozoa cannot be annulled by Act of Parliament," they state, in an obvious dig to women who were fighting for their right to vote. Their argument, stretched over more than three hundred pages including tables and line drawings of animals, explains how they see women as being complementary to men, like homemakers are to breadwinners, but certainly not able to achieve the same as them.

Another example is Darwin's cousin, the English scientist Francis Galton, remembered by history as the father of eugenics and for his devotion to measuring the differences between people. Among his quirkier projects was a "beauty map" of Britain, produced near the end of the nineteenth century by secretly watching local women and grading them from the ugliest to the most attractive. Brandishing their rulers and microscopes, men like Galton hardened sexism into something that couldn't even be challenged. Being able to gauge and standardize coated what would otherwise have been seen as ridiculous enterprises with the sweet perfume of scientific respectability.

Taking on this male scientific establishment wasn't easy, of course. But for Victorian women—women like Caroline Kennard—everything was at stake. They were fighting for their fundamental rights. They weren't even recognized as full citizens by their own countries. By 1887 only two-thirds of US states allowed a married woman to keep her own earnings. And it wasn't until 1882 that married women in the United Kingdom were allowed to own and control property in their own right.

Kennard and others in the women's movement realized that the intellectual debate over the inferiority of women could only be won on intellectual grounds. Like the male biologists attacking them, they would also have to deploy science to defend themselves. English writer Mary Wollstonecraft, who lived a century earlier, urged women to educate themselves. "Till women are more rationally educated, the progress of human virtue and improvement in knowledge must receive continual checks," she wrote in *A Vindication of the Rights of Woman* in 1792. Prominent Victorian suffragists made similar arguments, using what education they were allowed to have to question what was being written about women.

The new and controversial science of evolutionary biology became a particular target. Antoinette Brown Blackwell, believed to be the first woman ordained by an established Protestant denomination in the United States, complained that Darwin had neglected sex and gender issues. Meanwhile Charlotte Perkins Gilman, who authored the feminist short story "The Yellow Wallpaper," turned Darwinism around to argue for reform. She thought that half the human race had been kept down at a lower stage of evolution by the other half. With equality, women would finally have the chance to prove themselves equal to men. She was ahead of her time in many ways, arguing against a stereotyped division of toys for boys and girls and foreseeing how a growing army of working women might change society in the future.

But one Victorian thinker took on Darwin on his own turf, writing a book that passionately and persuasively argued on scientific grounds that women were not inferior to men.

> *"It seemed clear to me that the history of the life on the earth presents an unbroken chain of evidence going to prove the importance of the female."*

Unconventional ideas can appear from anywhere, even the most conventional of places.

The township of Concord in Michigan is one of those places. Home to fewer than three thousand people, it's an almost entirely white corner of America. The area's biggest attraction is a preserved post–Civil War house covered in pale clapboard siding. In 1894, not long after this house was built, a middle-aged schoolteacher from right here in Concord published some of the most radical ideas of her age. Her name was Eliza Burt Gamble.

We don't know much about Gamble's personal life, except that she was a woman who had no choice but to be independent. She had lost her father when she was two, her mother when she was around sixteen. Left without support, she made a living by teaching at local public schools. According to some reports, she went on to achieve impressive heights in her career. She also married and had three children, two of whom died before the century was out. Gamble's life could have been mapped out for her, the way it was for most middle-class women. She could have been a quiet, submissive housewife of the kind celebrated by the poet Coventry Patmore. Instead, she joined the growing suffrage movement to fight for the equal rights of women, becoming one of the most important campaigners in her region. In 1876 she organized the first women's suffrage conference in her home state of Michigan.

Gamble believed there was more to the cause than securing legal equality. One of the biggest sticking points in the fight for women's rights, she recognized, was that society had come to believe women were built to be lesser than men. Convinced this was wrong, in 1885 she set out to find hard proof for herself. She spent a year studying the collections at the Library of Congress, scouring the books for evidence. She was driven, she wrote, "with no special object in view other than a desire for information."

Evolutionary theory, despite what Charles Darwin had written about women, actually offered great promise to the women's movement. It opened a door to a revolutionary new way to understand humans. "It meant a way to be modern," says historian Kimberly Hamlin, whose 2014 book *From Eve to Evolution: Darwin, Science, and Women's Rights in Gilded Age America* charts women's responses to Darwin. Evolution was an alternative to religious stories that painted woman as man's spare rib. Christian models for female behavior and virtue were challenged. "Darwin created a space where women could say that maybe the Garden of Eden didn't happen . . . and this was huge. You cannot overestimate how important Adam and Eve were in terms of constraining and shaping people's ideas about women."

Although not a scientist herself, through Darwin's work Gamble realized just how devastating the scientific method could be. If humans were descended from lesser creatures, the same as all other life on earth, then it made no sense for women to be confined to the home or subservient to men. These obviously weren't the rules in the rest of the animal kingdom.

"It would be unnatural for women to sit around and be totally dependent on men," Hamlin tells me. The story of women could be rewritten.

In reality of course, for all the latent revolutionary power in his ideas, Darwin himself never believed that women were the intellectual equals of men. And this wasn't just a disappointment to Gamble but, judging from her writing, a source of great anger. She believed that Darwin, though correct in concluding that humans evolved like every other living thing on earth, was clearly wrong when it came to the role that women had played in human evolution.

Her criticisms were passionately laid out in a book she published in 1894 called *The Evolution of Woman, an Inquiry into the Dogma of Her Inferiority to Man*. "It was shocking," says Hamlin. Marshalling history, statistics, and science, this was Gamble's piercing counterargument to Darwin and other evolutionary biologists. She angrily tweezed out their inconsistencies and double standards. The peacock might have had the bigger feathers, she argued, but the peahen still had to exercise her faculties in choosing the best mate. And on the one hand, Darwin suggested that gorillas were too big and strong to become higher social creatures like humans. Yet at the same time he used the fact that men are on average physically bigger than women as evidence of their superiority.

He had also failed to notice, Gamble wrote, that the human qualities associated more commonly with women—cooperation, nurture, protectiveness, egalitarianism, and altruism—must have played a vital role in human progress. In evolutionary terms, drawing assumptions about women's abilities from the way they happened to be treated by society at that moment was narrow-minded and dangerous. Women had been systematically suppressed over the course of human history by men and their power structures, Gamble argued. They weren't naturally inferior. They just seemed that way because they hadn't been allowed the chance to develop their talents.

Gamble suggested that Darwin hadn't accounted for the existence of powerful women in some tribal societies either, which might prove that the supremacy of men now was not how it had always been. The ancient Hindu text the Mahabharata, which she picks out as an example, spoke of women being unconfined and independent before marriage was invented. So she couldn't help but wonder if "the law of equal transmission" applied to men as well as women: might it not be possible that males had been dragged along by the superior female of the species?

"When a man and woman are put into competition, both possessed of every mental quality in equal perfection, save that one has higher energy, more patience and a somewhat greater degree of physical courage, while the other has superior powers of intuition, finer and more rapid perceptions and a greater degree of endurance, . . . the chances of the latter for gaining the ascendancy will doubtless be equal to those of the former," she argues.

————

Eliza Burt Gamble's message, like that of other scientific suffragists, proved popular. Their provocative message was that women had been cheated out of the lives they deserved, that equality was in fact their biological right. "It seemed clear to me that the history of the life on the earth presents an unbroken chain of evidence going to prove the importance of the female," Gamble writes in the preface to the revised edition, which came out in 1916.

But even an army of readers and the support of fellow activists couldn't help win biologists around to her point of view. Her arguments were doomed to never fully enter the scientific mainstream, only circulate outside it.

But she never gave up. She marched on in her campaign for women's rights and continued writing for the press. Fortunately, she lived just long enough to see her own work as well as that of the wider movement gain real strength. In 1893 New Zealand became the first self-governing country to grant women the vote. The battle would take until 1918 in Britain, although only for women over the age of thirty. And when Gamble died in Detroit in 1920, it was just a month after the United States ratified the Nineteenth Amendment, which prohibited citizens from being denied the right to vote because of their sex.

While the political battle was a success, the war to change people's minds was taking much longer. "Gamble's ideas were praised in reform magazines and her writing style was generally praised, but the scientific and main-stream press balked at her conclusions and at her pretensions to write about 'science,'" says Hamlin. *The Evolution of Woman* was quite widely reviewed in newspapers and academic journals, but scarcely left a dent on science. "They were just like, 'Those silly women and their silly ideas.'"

A scathing book review in the *American Journal of Sociology* in 1915 reveals just how desperately some scientists clung to their prejudices, even when society around them was changing. "It must have been a sense of humor which led the publishers to put this volume in their 'Science Series,'" wrote the Texas University sociologist and liberal thinker Albert Wolfe about *Sex*

Antagonism, the latest work of the respected British biologist Walter Heape. Heape had taken his considerable scientific knowledge of reproductive biology and applied it somewhat less objectively to society, arguing that equality between the sexes was impossible because men and women were built for different roles.

Many biologists at the time agreed with Heape, including Scottish naturalist and coauthor of *The Evolution of Sex*, John Arthur Thompson, who gave the book a positive review. But Wolfe immediately saw the danger in scientists like him overstepping their expertise. "It is a fine illustration of the sort of mental pathology a scientist, especially a biologist, can exhibit when, with slight acquaintance with other fields than his own, he ventures to dictate from 'natural law' (with which Mr Heape claims to be in most intimate acquaintance) what social and ethical relation shall be," Wolfe mocked in his review. "He sees only disaster and perversion in the modern woman movement."

Parts of science remained doggedly slow to change. Evolutionary theory progressed pretty much as always, learning few lessons from critics such as Albert Wolfe, Caroline Kennard, and Eliza Burt Gamble. It's hard to picture the directions in which science might have gone if in those important days when Charles Darwin developed his theories of evolution, society hadn't been as sexist as it was. We can only imagine how different our understanding of women might be now if Gamble had been taken more seriously. Historians today have regrettably described her radical perspective as the road not taken.

In the century after Gamble's death, researchers became only more obsessed by sex differences, how they might pick them out, measure and catalogue them, enforcing the dogma that men are somehow better than women.

"Finding gold in the urine of pregnant mares."

It's perhaps appropriate that one of the next breakthroughs in the science of sex differences came courtesy of a castrated cock.

In the 1920s a fresh string of discoveries in Europe would alter the way science understood the differences between women and men just as much as Charles Darwin and evolutionary theory had. They were foreshadowed by a strange experiment in 1849, carried out by a German medical professor, Arnold Adolph Berthold. He had been studying castrated cockerels, commonly known as capons. It was known that by removing their testes, these birds were left with deliciously tender meat, which made them a popular delicacy at the time. Aside from their meat, live capons looked different from normal cocks. They were more docile. They could also be spotted by a characteristic red comb on top of their heads and unusually droopy red wattles.

The question for Berthold was, why?

He took the testes from normal cockerels and transplanted them into capons to see what happened. Remarkably, he found the capons started to look and sound like cocks again. The testes were surviving inside them, and growing. It was a startling result, but still nobody at the time understood why. What was it in the testes helping the capons seemingly come back from castration?

Progress came slowly. In 1891 another unusual experiment, this time in France by university professor Charles-Édouard Brown-Séquard, finally began to get to the root of the mystery. He suspected that male testes might contain some unknown substance that influenced masculinity. He proved his hypothesis the hard and fast way, by repeatedly injecting himself with a concoction made out of the blood, semen, and juices from the crushed testicles of guinea pigs and dogs. He claimed (although his findings were never replicated) that this cocktail increased his strength, stamina, and mental clarity.

The *British Medical Journal* reported Brown-Séquard's findings with excitement, describing the substance he had found as the "pentacle of rejuvenescence." Later, researchers carrying out similar experiments using female juices from guinea pig ovaries claimed to see a parallel feminizing effect. Over time, the secret juices inside all these male and female gonads were understood to be a specific set of chemicals, named "hormones."

We now know that sex hormones, found in the gonads, are just a handful of the fifty or more hormones produced across the human body. We can't live without them. They are the grease to our wheels. They've been described as "chemical messengers," delivering memos throughout the body to make sure it does the things it's supposed to do, including growing and keeping a stable temperature. From insulin to thyroxine, they helpfully regulate the functions of all sorts of organs. The sex hormones in particular regulate sexual development and reproduction. The two main female ones are estrogen and progesterone. Estrogen is what causes a woman's breasts to develop, among other things, while progesterone helps her body prepare for pregnancy. Male sex hormones are known as "androgens," of which the most well known is testosterone.

Sex hormones play a crucial role in determining how male or female a person looks even before birth. In the womb, it's interesting to note, all fetuses start out physically female. "The default blueprint is female," says Richard Quinton, consultant endocrinologist at hospitals in Newcastle-upon-Tyne in northeast England. About seven weeks after the egg has been fertilized, testosterone produced by the testes begins physically turning the male fetuses into boys. "Testosterone says: 'Make me externally male.'" Meanwhile another hormone stops this freshly male fetus from growing a uterus, fallopian tubes, and other female parts. As we grow older, hormones again play a role in puberty and beyond.

It's not surprising, then, that the discovery of sex hormones was one of the most important milestones in understanding what it means to be a woman or a man.

———

According to work done by social researcher Nelly Oudshoorn, now based at the University of Twente, in the Netherlands, hormone research sent waves of excitement through the pharmaceutical industry in the 1920s. Suddenly here was a way of scientifically understanding masculinity and femininity. With some effort, drug companies believed they could isolate and industrialize the production of sex hormones to make people more masculine or feminine.

By the early twentieth century, endocrinology—the new and controversial study of hormones—was turning into big business. Tons of animal ovaries and testes were harvested and thousands of liters of horse urine were collected as scientists desperately searched for chemicals that defined what it meant to be male or female. The director of Dutch pharmaceutical company Organon described it as "finding gold in the urine of pregnant mares."

By the end of the 1920s, treatments based on sex hormones were becoming available, and there was no limit to what they promised. In the archives of London's Wellcome Library, which keeps an enormous trove of historical medical documents, I find an advertising pamphlet from around 1929, produced by the Middlesex Laboratory of Glandular Research in London. It proudly announces that it's finally possible to replenish the "fire of life," to cure impotence, frigidity, and sterility using "the therapeutic utilisation of the sex hormones of fresh glands removed from healthy animals, such as the bullock, ram, stallion, ape." Treatments containing estrogen made similar claims aimed at women, guaranteeing to cure irregular periods and symptoms of menopause.

Of course, hormone treatments couldn't possibly live up to all this hype. But they weren't a fad either. They really did seem to work for certain symptoms, even if this was only anecdotally. An article in the *Lancet* medical journal in 1930 describes a male patient given testosterone, saying that he thought "his muscles were firmer and he felt more pugnacious; he nearly had a fight with his workmate." Another man, age sixty, was able "to play thirty-six holes of golf in a day without undue fatigue." Testosterone became associated with what were believed to be manly qualities, such as aggression, physical power, high intellect, and virility.

The same research was done on women using estrogen. Another article in the *Lancet* in 1931, the researcher Jane Katherine Seymour has noted, connected the female hormones to femininity and childbearing. Under their effect, it also said, women "would tend to develop a more passive and emotional, and less rational, attitude towards life."

In the early days of endocrinology, assumptions about what it meant to be masculine or feminine came from the Victorians. With the discovery of hormones, scientists had a new way to explain the stereotypes. According to Anne Fausto-Sterling, professor of biology and gender studies at Brown University, the prominent British gynecologist William Blair-Bell, for instance, believed that a woman's psychology depended on the "state of her internal secretions" keeping her in "her normal sphere of action." At that time, this meant being a wife and mother. If she stepped outside these social boundaries, scientists like him implied it must be because her hormone levels were out of whack.

Researchers thought that sex hormones were doing more than just affecting reproductive behavior. They were also responsible for making men manlier, by the standards of the time, and for making women womanlier, again by the standards of the time. Reasoning in this way, scientists assumed the sex hormones belonged uniquely to each sex. Male hormones—androgens—could only be produced by men, and female hormones—estrogen and progesterone—could only be produced by women. After all, if they were the key to manliness and womanliness, why would it be any other way?

An interesting experiment in 1921 hinted at the possibility that all the assumptions that scientists were making about sex hormones might be wrong.

A Viennese gynecologist revealed that treating a female rabbit with an extract from an animal's testes changed the size of her ovaries. Later, to their shock, scientists began to realize that significant levels of androgens were present in women and of estrogen in men. In 1934, the German-born gynecologist Bernhard Zondek, while studying stallion urine, reported on "the paradox that the male sex is recognised by a high oestrogenic hormone content." In fact, a male horse's testes turned out to be one of the richest sources of estrogen ever found.

Just when endocrinologists thought they were getting a grip on what the sex hormones did, this threw everything into confusion. And it raised an interesting dilemma: If estrogen and testosterone determined femaleness and maleness, then why did both sexes naturally have both? What, then, did it even mean to be born male or female?

For a while, some scientists thought that female sex hormones might be turning up in men because they had eaten them. This bizarre food hypothesis was ditched when it gradually became clear that male and female gonads can in fact produce both hormones themselves. Others then thought that the only thing that estrogen could be doing in a man was pulling him away from masculinity and toward femininity, perhaps even toward homosexuality.

It took a while for scientists to accept the truth: that all these hormones really did work together in both sexes, in synergy. Oudshoorn has described how important a shift this was in the way that science understood the sexes. Suddenly a spectrum opened up on which men could be more feminine and women more masculine, instead of opposites. Writing in 1939 at the end of what he described as this "epoch of confusion," Herbert Evans at the Institute of Experimental Biology at the University of California, Berkeley, admitted, "It would appear that maleness or femaleness can not be looked upon as implying the presence of one hormone and the absence of the other. . . . Though much has been learned it is only fair to state that these differences are still incompletely known."

The implications of this change of thinking were spectacular. The entire notion of what it meant to be a woman or a man was up for grabs. Researchers in other fields began to explore the boundaries of sexual and gender identity. Cultural anthropologist Margaret Mead, for example, started writing at about the same time about masculine and feminine personalities, and how culture rather than biology might be influencing which ones people had. Studying Samoan communities in 1949, she wrote, "The Samoan boy is not over pressured into displays of manhood, and the girl who is ambitious and managing has plenty of outlets in the bustling, organised life of the women's groups." The Mundugumor tribe of Papua New Guinea, she also noticed, created women with a more typically male temperament.

Not everyone today agrees with Mead's observations, but her ideas did signal how society was changing, in part prompted by science. A radical move from the old Victorian orthodoxies of the kind Charles Darwin had subscribed to was underway. People could no longer clearly define the sexes anymore. There was overlap. Femaleness and maleness, femininity and masculinity, were turning into fluid descriptions, which might be as much shaped by nurture as by nature.

This revolution in scientific notions of what it meant to be a woman also came in time for the second wave of feminism in the 1960s and 1970s, following the pioneering movement decades earlier that had earned women the vote. By now, female biologists, anthropologists, and psychologists were entering universities and graduating in growing numbers. They were becoming researchers and professors. This helped research on women to enter another era. Fresh ideas challenged long-standing narratives.

The path paved by Eliza Burt Gamble, the pioneering suffragist who had dared to challenge Charles Darwin more than a hundred years before, was being trodden once again by a new generation of scientists.

———————

We arrive at today.

Lingering stereotypes about sex hormones remain. But they are being constantly challenged by new evidence. According to endocrinologist Richard Quinton, common assumptions about testosterone have already been shown to be way off the mark. Women with slightly higher than usual levels of testosterone, he says, "don't actually feel or appear any less feminine."

In 2008, former Wall Street trader John Coates, a neuroscientist at Cambridge University who researches the biology of risk taking and stress, decided to see whether the cliché of stock market trading floors being testosterone-fueled dens of masculinity was true. He took saliva samples from traders and found that when their testosterone levels were above average, their gains were also above average. Another study in 2015 by a large team of scientists across the United Kingdom, United States, and Spain revealed that testosterone wouldn't have made the traders any more aggressive. It just made them slightly more optimistic. And when it came to future price changes, this may have helped them take a few more risks.

Quinton similarly claims to have seen no link between testosterone and aggression among his own patients, despite the stereotype that it makes people more violent. "I'm not sure where it comes from," he tells me. "Urban myth?"

The balance between nature and nurture is starting to be a little better understood. In academic circles at least, gender and sex are now recognized as two different things. Sex is something scientifically distinct for most people. It's defined by a certain package of genes and hormones as well as more obvious physical features, including a person's genitals and gonads (although a small proportion of people are biologically intersex). Gender, meanwhile, is a social identity, influenced not only by biology but also by external factors such as upbringing, culture, and the effect of stereotypes. It's defined by what the world tells us is masculine or feminine, and this makes it potentially fluid. For many people their biological sex and their gender aren't the same.

But we remain in the early days of this research. The biggest questions are still unanswered. Does the balance of sex hormones have an effect beyond the sexual organs and deeper into our minds and behavior, leading to pronounced differences between women and men? And what does this tell us about how we evolved? Is the traditional stereotype of the breadwinning father and the stay-at-home mother really part of our biological makeup, as Darwin assumed, or is it an elaborate social construction that's unique to humans? Studies into sex differences are as powerful as they are controversial. The same way that research on hormones challenged popular wisdom about masculinity and femininity in the twentieth century, science is forcing us to question all aspects of ourselves.

The facts, as they emerge, are important. In a world in which so many women continue to suffer sexism, inequality, and violence, they are capable of turning old stereotypes on their head. They can transform the way we see each other. With good research and reliable data, the strong can become weak and the weak, strong.

CHAPTER 2

Females Get Sicker but Males Die Quicker

The evidence is clear: from the constitutional standpoint woman is the stronger sex.

—Ashley Montagu,
The Natural Superiority of Women, 1953

"It's wonderful," says Mitu Khurana, a hospital administrator living in New Delhi, India. "When you have your first pregnancy, everyone is very excited. It is a feeling beyond description."

The moment she's so fondly remembering was a decade ago. She had become pregnant with twins just a few months after getting married, and she assumed that nothing could ruin her happiness. Raised in a family of sisters, Khurana didn't care whether she was having boys or girls, or one of each. "I just wanted the children to be healthy," she tells me.

But her husband and his family didn't feel the same way. They wanted sons.

So begins a common story. It's one that has been repeated in millions of homes across India, China, and other parts of South Asia, where cultures unashamedly prize sons above daughters. They are cultures, as Khurana learned all those years ago, that will go to terrible lengths to stop a girl from being born. Some women keep having children until they finally have a boy. Others are pressured to abort female fetuses, even to the point of torture. If they do make it to the day of their birth, many babies and young girls are routinely treated worse than boys. In the most appalling cases, they are killed. In 2007 police in Orissa in eastern India found skulls and body parts of what they believed to be three dozen female fetuses and infants down a disused well. One 2013 news report described a baby buried alive in a forest in the central state of Madhya Pradesh. Another in 2014 told of a newborn in Bhopal dumped in a rubbish bin.

That year a United Nations report described the problem as having reached emergency levels. India's 2011 census had already revealed that there were more than seven million fewer girls than boys age six and under. The overall sex ratio was more skewed in favor of boys than it had been a decade ago. One reason for this worsening in the records was the growing availability of prenatal scans, which for the first time allowed parents to find out the sexes of their babies easily and early enough to have selective abortions.

In 1994 the Indian government outlawed sex selection tests, but unscrupulous independent clinics and doctors still offer them for a fee, in private and under the radar. Khurana never wanted to have one of these prenatal scans, she tells me. In the end, she wasn't given the choice. During her pregnancy, she claims she was tricked into eating some cake that contained egg, to which she's allergic. Her husband, a doctor, then took her to a hospital, where a gynecologist advised her to have a kidney scan under sedation. It was then, she believes, he deliberately found out the sex of her babies without her consent or knowledge.

"I knew it from his behavior that I'm getting daughters," she explains. He and his family immediately began pushing her for an abortion. "There was a lot of pressure." She says she was denied food and water and once pushed down the stairs. Desperate and frightened, Khurana went to stay with her parents and eventually gave birth to her daughters while she was there.

She managed to save her girls. But things didn't change. "They were not at all warm," she recalls of her husband and his family's attitude toward her daughters. A few years later she stumbled on an old hospital report revealing the sex of her fetuses. She read it as proof that her husband had indeed carried out an ultrasound scan on her while she was pregnant. And that discovery helped launch a legal case against both him and the clinic, which is still making its way through the notoriously slow Indian courts by the time I interview her, ten years since the birth of her daughters. Her husband and the clinic have both strongly denied her allegations in the press.

Now long separated from her husband and awaiting a divorce, Khurana has become famous in India for being among the first women to take this kind of legal action. Taking her campaign across the country has also confirmed just how widespread a problem this is, blind to class or religion. "I'm fighting because I don't want my daughters to go through this. Women are wanted as wives and girlfriends but not as daughters," she says. "What my husband did was part of social conditioning. I don't blame him anymore. He's a by-product of society, and society has to change."

However well hidden the selective abortions, murders, and abuse of mothers and their girls, the countrywide statistics don't lie. Reality is laid bare in the grotesquely uneven sex ratios. The United Nations report *The World's Women 2015* says, "For those countries in which the sex ratio falls close to or below the parity line, it can be assumed that discrimination against girls exists."

It is a situation familiar to Joy Lawn, director of the Centre for Maternal, Adolescent, Reproductive, and Child Health at the London School of Hygiene and Tropical Medicine. "You go to hospitals in South Asia and there can be whole wards of kids with illnesses, and you will find 80 percent of them are boys because the girls aren't being brought to the hospital," she tells me. A similar gender imbalance was uncovered in a 2002 study in Nepal, northeast of India, by public health researchers Miki Yamanaka and Ann Ashworth, also from the London School of Hygiene and Tropical Medicine. They looked at how much work children are expected to do to support their families and found that girls worked twice as long as boys and that their work was also more physically demanding.

The effects that society can have on gender differences are profound, but none quite so profound as the taking of life altogether. What makes the mortality figures even more shocking is that, contrary to assumptions about women being the weaker sex, a baby girl is statistically more robust than a baby boy. She's naturally better built to live. As scientists start to explore the female body in truer detail, they are learning just how powerful a girl's survival edge is—even in a world that doesn't always want her.

"Pretty much at every age, women seem to survive better than men."

We often think of males as being the tougher and more powerful sex. It's true that men are on average six inches taller than women and have around double the upper body strength. But then, strength can be defined in different ways. When it comes to the most basic instinct of all—survival—women's bodies tend to be better equipped than men's.

The difference exists from the moment a child is born.

"When we were there on the neonatal unit and a boy came out, you were taught that, statistically, the boy is more likely to die," explains Lawn. Besides her academic research into child health, she has worked in neonatal medicine in the United Kingdom and as a pediatrician in Ghana. The first month following birth is the window in which humans are at their greatest risk of death. A million babies die on the day of their birth every year. But if they receive exactly the same level of care, females are statistically less likely to die than males. Lawn's research encompasses data from across the globe, giving the broadest picture possible of infant mortality. And having researched the issue in such depth, Lawn concludes that boys are at around a 10 percent greater risk than girls in that first month—and this is at least partly, if not wholly, for biological reasons.

In South Asia then, the mortality figures should be in favor of girls. That they're not even equal but so far skewed in favor of boys means that girls' natural power to survive is being forcibly degraded by the societies they are born into. "If you have parity in your survival rates, it means you aren't looking after girls," says Lawn. "The biological risk is against the boy, but the social risk is against the girl."

Elsewhere, child mortality statistics bear this out. For every thousand live births in sub-Saharan Africa, ninety-eight boys compared with eighty-six girls die by the age of five. A study that Lawn and her colleagues published in the journal *Pediatric Research* in 2013 confirmed that a boy is 14 percent more likely to be born prematurely and more likely to suffer disabilities, ranging from blindness and deafness to cerebral palsy, when he's at the same stage of prematurity as a girl. In the same journal in 2012 a team at King's College London reported that male babies born very prematurely are more likely to stay longer in the hospital, die, or suffer brain and breathing problems.

"I always thought that it was physically mediated, because boys are slightly bigger, but I think it's also biological susceptibility to injury," adds Lawn. One explanation for more boys being born preterm is that mothers expecting boys are, for reasons unknown, more likely to have placental problems and high blood pressure. Research published by scientists at the University of Adelaide in the journal *Molecular Human Reproduction* in 2014 showed that newborn girls may be healthier on average because a mother's placenta behaves differently depending on the sex of the baby. With female fetuses, the placenta does more to maintain the pregnancy and increase immunity against infections. Why this is, nobody understands. It could be because, before birth, the normal human sex ratio is slightly more skewed toward boys. The difference after birth might simply be nature's way of correcting the sex balance.

But the reasons could also be more complicated. After all, a baby girl's natural survival edge stays with her throughout her entire life. Girls aren't just born survivors, they grow up to be better survivors, too.

"Pretty much at every age, women seem to survive better than men," confirms Steven Austad, international expert on aging and chair of the biology department at the University of Alabama, Birmingham. He describes women as being more "robust." It's a phenomenon so clear and undeniable that some scientists believe understanding it may even hold the key to human longevity.

At the turn of the millennium, Austad began to investigate exactly what it is that helps women outlive men at all stages of life. "I wondered if this is a recent phenomenon. Is this something that's only true in industrialized countries in the twentieth century and twenty-first century?" Digging through the Human Mortality Database, a collection of longevity records from around the world and founded by German and American researchers in 2000, he was surprised to discover that the phenomenon really does transcend time and place.

The database now covers thirty-eight countries and areas. But his favorite example is Sweden, which has kept some of the most thorough and reliable demographic data anywhere. In 1800 life expectancy at birth in Sweden stood at thirty-three years for women and thirty-one for men. In 2015 it was about eighty-three years for women and about seventy-nine for men. "Women are more robust than men. I think there's little doubt about that," he says. "It was true in the eighteenth century in Sweden, and it's true in the twenty-first century in Bangladesh, and in Europe, and in the US."

I ask Austad whether women might be naturally outliving men for social reasons. It's reasonable to think, for instance, that boys are generally handled more roughly than girls are. Or that more men than women take on risky jobs, such as construction and mining, which also expose them to toxic environments. And we know that in total across the world, far more men than women smoke, and that this habit dramatically pushes up their mortality rates. But Austad is convinced that the difference is so pronounced, ubiquitous, and timeless that it must mean features in a woman's body underlie the difference. "It's hard for me to imagine that it is environmental, to tell you the truth," he says.

The picture of this survival advantage is starkest at the end of life. The Gerontology Research Group keeps a list online of all the people in the world who they have confirmed are living past the age of 110. I last checked the site in July 2016. Of all these "supercentenarians" in their catalogue, just two were men. Forty-six were women.

Yet we don't know why.

––––––––––

"I'm absolutely puzzled by it," says Steven Austad. "When I first started looking into this, I expected to find a huge literature on it, and I found virtually nothing. There's a big literature on 'Is this a difference between men and women?' but the underlying biology of the survival difference, there's very little on that. It's one of the most robust features of human biology that we know about, and yet it's had so little investigation."

For more than a century, scientists have painstakingly studied our anatomy, even collected thousands of liters of horse urine to root out the chemicals that make men more masculine and women more feminine. Their search for sex differences has shown no boundaries. But when it comes to why women might be more physically robust than men—why they are better survivors—research has been scarce. Even now, only scraps of work here and there point to answers.

"It's a basic fact of biology," observes Kathryn Sandberg, director of the Center for the Study of Sex Differences in Health, Aging and Disease at Georgetown University, who has explored how much of a role disease has to play in why women survive. "Women live about five or six years longer than men across almost every society and that's been true for centuries," she explains.

"First of all, you have differences in the age of onset of disease. So, for example, cardiovascular disease occurs much earlier in men than women. The age of onset of hypertension, which is high blood pressure, also occurs much earlier in men than women. There's also a sex difference in the rate of progression of disease. If you take chronic kidney disease, the rate of progression is more rapid in men than in women." Even in laboratory studies on animals, including mice and dogs, females have done better than males, she adds.

By picking through the data, researchers like Kathryn Sandberg, Joy Lawn, and Steven Austad have come to understand just how widespread these gaps are. "I assumed that these sex differences were just a product of modern westernized society, or largely driven by the differences in cardiovascular diseases," says Austad. "Once I started investigating, I found that women had resistance to almost all the major causes of death." One of his papers shows that in the United States in 2010, women died at lower rates than men from twelve of the fifteen most common causes of death, including cancer and heart disease, when adjusted for age. Of the three exceptions, their likelihood of dying from Parkinson's or stroke was about the same. And they were more likely than men to die of Alzheimer's disease.

When it comes to fighting off infections from viruses and bacteria, women also seem to fare better. "If there's a really bad infection, they survive better. If it's about the duration of the infection, women will respond faster, and the infection will be over faster in women than in men," says Sandberg. "If you look across all the different types of infections, women have a more robust immune response." It isn't that women don't get sick. They do. They just don't die from these sicknesses as easily or as quickly as men do.

One explanation for this gap is that higher levels of estrogen and progesterone in women might be protecting them in some way. These hormones don't just make the immune system stronger but also more flexible, according to Sabine Oertelt-Prigione, a researcher at the Institute of Gender in Medicine at the Charité university hospital in Berlin. "This is related to the fact that women can bear children," she explains. A pregnancy is the same as foreign tissue growing inside a woman's body that, if her immune system was in the wrong gear, would be rejected. "You need an immune system that's able to switch from proinflammatory reactions to anti-inflammatory reactions in order to avoid having an abortion pretty much every time you get pregnant. The immune system needs to have mechanisms that can, on one side, trigger all these cells to come together in one spot and attack whatever agent is making you sick. But then you also need to be able to stop this response when the agent is not there anymore, in order to prevent tissues and organs from being harmed."

The hormonal changes that affect a woman's immune system during pregnancy also take place on a smaller scale during her menstrual cycle, and for the same reasons. "Women have more plastic immune systems. They adapt in different ways," says Oertelt-Prigione. Many types of cells in the body are involved in immunity, but the kind that come into closest contact with viruses and bacteria are known as "T cells." T cells inject substances into bacteria to kill them or secrete other substances that call more cells to action, some of which "eat up" infected cells and bacteria, like Pac-Man in the video game, she explains. Researchers know that a certain type of T cell that's crucial to managing the body's response to infections becomes more active in the second half of a woman's menstrual cycle, when she's able to get pregnant.

The discovery that sex hormones and immunity might be linked is fairly recent. In men, scientists have explored connections between testosterone and lower immunity, although the evidence is relatively thin. In 2014 Stanford University researchers found that males with the highest levels of testosterone, for example, had the lowest antibody response to a flu vaccine, which meant they were the least likely to be protected by the jab. As yet, though, it's an unsubstantiated link. In women, the connection is far clearer; so much so that patients themselves have noticed these fluctuations. For years, doctors assumed that a woman's immunity couldn't be changing during her menstrual cycle. If she did report a difference in pain levels, doctors might dismiss it as premenstrual syndrome or some vague psychological complaint. It was only when these links were increasingly backed up by hard research that scientific interest was sparked and more research began to flourish.

This problem runs all the way through research into women's health. If a phenomenon affects women and only women, it's all too often misunderstood. And this is compounded by the fact that even though they're good at surviving, women aren't healthier than men. In fact, quite the opposite.

"If you could add up all the pain in the world, all the physical pain, I suspect that women have way, way more of it. This is one of the penalties of being a better survivor. You survive, but maybe not quite as intact as you were before," says Steven Austad. Statistically, it could even explain why women seem proportionally sicker than men. "Part of the reason that there are more women than men around in ill health is to do with the fact that women have survived events that would kill men, and so the equivalent men are no longer with us."

Another reason is that women's immune systems are so powerful that they can sometimes backfire. "You start regarding yourself as foreign and your immune system starts attacking its own cells," explains Kathryn Sandberg. Diseases caused in this way are known as "autoimmune disorders," the most common of which include rheumatoid arthritis, lupus, and multiple sclerosis. "It's kind of a double-edged sword with the immune system. In some ways it's better to have a female immune system if you're fighting off infection of any kind, but on the other hand, we are more susceptible to autoimmune diseases, which are very problematic," she explains.

This isn't to say autoimmune disease is always hardest on women. When men develop multiple sclerosis, they tend to get it worse. Women also survive with it longer than men do. Even so, of the roughly 8 percent of Americans who suffer from autoimmune diseases, estimates suggest that at least three-quarters are women.

"In autoimmune diseases, they almost all tend to get worse right before the menstrual cycle in women who are premenopausal," says Sabine Oertelt-Prigione. There are theories postulating that the same way varying hormone levels may boost a woman's immunity at different times of the month, they might also affect her experience of illness. According to some reports, for instance, women with asthma are at highest risk of an attack just before or at the start of their period. When estrogen and progesterone levels drop through the years following the start of menopause, a woman's immunity advantage starts to drop away as well.

When it comes to viral infections, too, a woman's strong immune response may be a problem as well as a benefit. Research on influenza by Sabra Klein, an immunologist at the Johns Hopkins Bloomberg School of Public Health, has shown that while women are generally hit by fewer viruses during an infection, they tend to suffer more severe flu symptoms than men do. She reasons that this may be because women's immune systems mount sturdier counterattacks to viruses, but then suffer when the effects of these counterattacks upset their own bodies.

Women also tend to get more painful joint and muscle diseases in general, observes Austad. Part of this is due to autoimmune diseases that affect the joints, such as arthritis. The physical toll of childbearing and the hormonal changes of menopause may also leave women with more physical problems and disabilities, especially in later life. Bone density is known to fall in the short term after pregnancy and also after menopause. Weight gain is now, too, recognized as a symptom of menopause.

But the overall picture of pain and ill health is complicated. "Cross-culturally women just report more physical limitations and more disabilities. It's really widespread," says Austad. When it comes to biological clues about the underlying reasons for this sex difference in disease or survival, however, he adds, "I don't feel very confident of any explanation."

It's difficult to tear apart the effects of biology from other effects. Society and the environment can sometimes affect illness more than a person's underlying biology. "Women are less likely to go to the hospital when they're feeling chest pain than men," explains Sandberg, who has looked at gender differences in heart disease in particular. Men's and women's health habits throughout the world differ in countless other ways. Oertelt-Prigione points out that where families eat collectively and food is scarce, women are sometimes the last to eat and the most likely to give up food, which can raise their risk of malnutrition. This in turn can affect their susceptibility to disease.

Not only a woman's own behavior but that of others around her can also affect her health. From the second a girl is born, she's placed in a different box. She may be handled differently, fed differently, and treated differently. This marks the beginning of a lifetime of differences in the way doctors and medical researchers approach her as well. Only recently, for instance, have doctors begun to acknowledge the severity of some women's experience of period pain. In 2016, a professor of reproductive health at University College London, John Guillebaud, told a reporter that it can be "almost as bad as having a heart attack," while admitting that period pain hasn't been given the attention it deserves, partly because men don't suffer from it. In 2015 a team of British researchers studying cancer diagnosis in the United Kingdom found that for six of the cancers that affect both men and women, including bladder and lung, it took longer for women to be diagnosed after going to doctors with their symptoms. For gastric cancer, a woman waited on average a full two weeks longer for a diagnosis.

If there *are* deep-seated biological sex differences in health, and the differences aren't largely due to society and culture, then scientists need to go deeper inside the body to find them.

———

"Females get sicker but males die quicker," says Arthur Arnold, a professor at the University of California, Los Angeles. It's an old adage bandied among his undergraduates. It reflects what doctors all over the world have observed, and Arnold is convinced this betrays the long roots of sex differences in health. He runs a laboratory studying the biological factors that make females different from males and edits the journal *Biology of Sex Differences*. His work has taken him beyond looking at organs and sex hormones and down to the fundamental level of the gene.

The human body is made up of trillions of cells. Every one of them has genetic information stored in chromosomes, which explain to the body how to build itself from the subtlest hormones all the way up to skin and bone. We have forty-six chromosomes in total, split into twenty-three pairs, and the roots of the genetic differences between men and women lie in our twenty-third pair, the "sex chromosomes." In women, these are called "XX,"

with one X chromosome inherited from each parent. Men's sex chromosomes are called "XY," with the X coming from the mother and the Y from
the father. For a long time scientists assumed that these sex chromosomes
were mainly concerned with reproduction and not much else. Today some,
including Arnold, believe that the consequences of this seemingly tiny genetic difference may stretch much farther.

Each chromosome in a pair carries the same genes in the same locations,
known as "alleles." For a female with two X chromosomes, the allele for eye
color from her father will be matched by another one for eye color in the
same place from her mother. For males with XY sex chromosomes, however,
a matching allele isn't always there. X and Y don't have the same genes in the
same locations. In fact, the Y is far smaller than the X.

Having just one copy of the genes, only on the X chromosome, can have
repercussions for a man's body. "It's long been thought, and there is good evidence for this, that having two versions of the gene buffers women against
certain diseases or environmental changes," says Arnold. If a man happens to
have a genetic mutation on one of his X chromosomes that causes an illness
or disability, he has no way of avoiding it. A woman, on the other hand, will
have an extra X chromosome to counteract it, unless she's unlucky enough
to have the same genetic mutation on both X chromosomes, one from each
parent. "The simple case would be if one gene works better when it's cold
and another works better when it's hot. A woman with both of those alleles
can be healthy when it's hot and cold. The male only gets one shot. He only
has one copy. So his body either works better when it's hot or works better
when it's cold, but not both."

Men are more susceptible to some well-known genetic traits simply because they have one X chromosome. They're known as "X-linked disorders."
They include red-green color blindness, hemophilia, muscular dystrophy,
and IPEX syndrome, which affects immune function. Mental retardation,
which affects 2 to 3 percent of people in developed countries and significantly more men than women, also has a strong link to the X chromosome.

This is one reason why, in the effort to understand sex differences in
health, Arnold has chosen to zero in on chromosomes. "We went back to
the most fundamental biological differences between males and females.
From the time of the fertilization of the egg, the only one thing that we
know is different between males and females is sex chromosomes. So everything has got to come from that, . . . everything's downstream of the sex
chromosomes."

"What we know of X-linked diseases is that they're pretty rare," says biologist Steven Austad. "But I think there are a lot more X-linked diseases than we think about. . . . This probably underlies a considerable proportion of the sex difference." One example is respiratory syncytial virus, which infects the lungs and breathing passages and is one of the biggest causes of bronchitis in children under the age of one in Britain and the United States. Researchers have found that the virus tends to hit boys far more than girls, and that something inside one particular gene on the X chromosome might be responsible.

Gender medicine researcher Sabine Oertelt-Prigione agrees that there may be genes linked to resilience, immunity, and disease susceptibility on the human X chromosome that haven't yet been discovered or understood. "In my school we were taught that the X and Y are basically related to sexual function. That's it. Nobody was thinking beyond that really at the time, and I'm talking about twenty years ago. Then things slowly started to change," she explains.

In 1961 English geneticist Mary Frances Lyon found that, even though women have two X chromosomes, one is randomly inactivated in every cell. In other words, only one of them shows up for work. Women are therefore a genetic mosaic in which some cells have genes from one X chromosome, and other cells have genes from the other. Researchers have more recently discovered that some genes on the second X chromosome aren't actually inactivated at all. Christine Disteche, a professor of pathology at the University of Washington, Seattle, and one of the world's leading researchers on X inactivation, describes them as "little islands of escape." In 2009 researchers at Penn State College of Medicine totted up these uninactivated genes to discover that these islands comprise 15 percent of genes on the second X. "We are now looking at huge data sets on gene expression between males and females, in humans and mice, to really try to see what is the extent of these differences," says Disteche.

"Finding out that one of the two is not completely inactivated, it leads to speculation about lots of interesting aspects of life for women. It may be the reason we live longer," suggests Oertelt-Prigione.

The problem for all researchers in this area is that it's not easy to distill the impact of the X chromosome from all the other factors that also cause a person to get sick or die. Most diseases don't appear to be linked to one or even a few genes in the way that X-linked genetic disorders such as hemophilia and muscular dystrophy are. The things that kill many of us, such as cardiovascular disease, are more complicated than that. Could genes from a second X chromosome have consequences for how the heart works, for instance?

To answer this question, biologist Arnold and his team have used a special laboratory animal, one with absolutely no difference between its males and females except for the number of X chromosomes it has. In nature, these creatures don't exist. But by using genetic modification, scientists can come close to building them. Since sex hormones before birth have the most obvious impact on male and female bodies (without androgens, a male wouldn't develop male gonads, for instance), researchers have created laboratory mice for Arnold that don't produce these hormones. The resulting mice have XY chromosomes, like a male, but also ovaries, like a female. This has allowed Arnold to compare genetically altered XY female mice to normal XX female mice. The only difference between them should be in their chromosomes. If their health differs, it's purely because of the effects of their genes.

The results have indeed shown a link between the number of X chromosomes a mouse has and its health. Arnold describes "three dramatic cases." When he and his team looked at body weight, they found that mice get fat if their gonads are removed. But animals with two X chromosomes get a lot fatter than those with just one. It mirrors something we see in human adults—women tend to have a higher percentage of fat mass in their bodies than men. "A second example is that if we give the mouse a heart attack, the animals with two X chromosomes do worse than the animals with one X chromosome," he adds. "And the third example in the mouse model is with multiple sclerosis, where we induced a multiple sclerosis–like disease in the mouse, and the animals that are XX do worse than the animals that are XY." Multiple sclerosis in humans, being an autoimmune disease, affects more women than men.

Their take-home message is that many of the sex differences we see in health are rooted deep down in genetics. "The study of mouse models has provided convincing evidence that cells with two X chromosomes are intrinsically different from those with one X chromosome. Sex differences caused by the number of X chromosomes can have a profound effect on disease," Arnold and his colleagues wrote in their paper about the experiment, published in 2016 in the journal *Philosophical Transactions of the Royal Society of London B*.

But not everyone is convinced. Some are dubious as to whether rodents can provide quite as much insight as Arnold believes they can. "Personally, I'm not a mouse fan," says Oertelt-Prigione. "I don't know how translatable findings in mice are to humans. . . . I think they have given us a lot of information, but I just wonder at this point how far we should pursue that."

Other criticisms are bigger. In her 2013 book, *Sex Itself: The Search for Male and Female in the Human Genome*, Harvard University professor of the social sciences Sarah Richardson questions the idea that every cell in the body is intrinsically different depending on someone's sex, and that this leads to the gaps we see between women and men. "It is a widely shared consensus among social scientists that genomics is transforming social relations," she writes, adding, "The same may be said of genetic research on sex and gender." Arnold, for instance, describes the effect of sex-biasing factors in our genes as a "sexome" (like the genome, but for sex difference). "You can think of the cell as this kind of big network," he tells me. "Males and females are different because they have different levels of sex-biasing factors, and they pull on the network at various points." This idea is that, even though the sex chromosomes are only one pair in twenty-three of all the pairs of chromosomes we have, their effects stretch much farther.

Richardson warns against this focus on genetics as an umbrella explanation for sex difference because of how it blurs away the effects of society and culture, as well as other biological factors. Age, weight, and race, for example, are known to have a huge impact on health. Hormones are important, too. She notes that the body of genetic evidence when it comes to sex differences paints an overwhelming picture of similarity. Indeed, Arnold himself admits to me that his idea of the sexome is "more of an evocative phrase" than a solid theory backed up by research.

The debate about just how deep the dividing line is between women and men continues to rage inside the scientific community. It has been fueled most recently by anger over exactly the opposite problem: the habit of medical researchers to leave women out of tests for new drugs, because their bodies were thought to be so similar to men's.

"It is much cheaper to study one sex."

"Let's face it, everyone in the biomedical community has spent their lives studying one sex or the other. And it's usually males," says biologist Steven Austad. When it comes to the basic machinery of our bodies, scientists have often assumed that studying one sex is as good as studying the other.

"I one time looked into the rodent literature on dietary restrictions," recounts Austad. "There are hundreds and hundreds of studies. And I found that there was only a handful that included both sexes. . . . People seem to be willing to extrapolate from one sex and just assume that everything they find is going to be true in the other sex."

In 2011 health researcher Annaliese Beery at the University of California, San Francisco, and biologist Irving Zucker at the University of California, Berkeley, published a study looking into sex biases in animal research in one sample year: 2009. Of the ten scientific fields they investigated, eight showed a male bias. In pharmacology, the study of medical drugs, the articles reporting only on males outnumbered those reporting only on females by five to one. In physiology, which explores how our bodies work, it was almost four to one.

It's an issue that runs through other corners of science, too. In research on the evolution of genitals (parts of the body we know for certain are different between the sexes), scientists have also leaned toward males. In 2014 biologists at Humboldt University in Berlin and Macquarie University in Sydney analyzed more than three hundred papers published between 1989 and 2013 that covered the evolution of genitalia. They found almost half looked only at the males of the species, while just 8 percent looked only at females. One reporter described it as "the case of the missing vaginas."

When it comes to health research, the issue is more complicated than simple bias. Until around 1990, it was common for medical trials to be carried out almost exclusively on men. And there were some good reasons for this. "You don't want to give the experimental drug to a pregnant woman, and you don't want to give the experimental drug to a woman who doesn't know she's pregnant but actually is," explains Arthur Arnold. The terrible legacy of women being given thalidomide for morning sickness in the 1950s proved to scientists how careful they need to be before giving drugs to expectant mothers. Thousands of children were born with disabilities before thalidomide was taken off the market.

"You take women of reproductive age off the table for the experiment, which takes out a huge chunk of them," continues Arnold. A woman's fluctuating hormone levels might also affect how she responds to a drug. Men's hormone levels are more consistent. "It is much cheaper to study one sex. So if you're going to choose one sex, most people avoid females because they have these messy hormones. . . . So people migrate to the study of males. In some disciplines it really is an embarrassing male bias," he adds.

This tendency to focus on males, researchers now realize, may have harmed women's health. "Although there were some reasons to avoid doing experiments on women, it had the unwanted effect of producing much more information about how to treat men than women," Arnold explains. A 2010 book on the progress in tackling women's health problems, cowritten by the Committee on Women's Health Research that advises the National Institutes of Health (NIH), notes that autoimmune diseases—which affect far more women than men—remain less well understood than some other conditions. "Despite their prevalence and morbidity, little progress has been made toward a better understanding of those conditions, identifying risk factors, or developing a cure," it states.

Another problem is that women may respond differently from men to certain drugs. Medical researchers in the mid-twentieth century often assumed this couldn't be a problem. "There was a notion that women were more like little men, . . . that if this treatment works in men, it will work on women," says Janine Clayton, director of the Office of Research on Women's Health at the NIH in Washington, DC, which funds the vast majority of US health research.

We now know this isn't necessarily true. In 2001 New Zealand–based dermatologist Marius Rademaker estimated that women are around one-and-a-half times as likely to develop an adverse reaction to a drug as men are. In 2000 the US Government Accountability Office looked at the ten prescription drugs withdrawn from the market since 1997 by the US Food and Drug Administration. Studying reported cases of adverse effects, it found that eight drugs posed greater health risks to women than to men. The withdrawn drugs included two appetite suppressants, two antihistamines, and one for diabetes. Four of these were simply given to many more women than men, but the other four showed these effects even when men took them in more equal numbers.

"You have to be concerned that there were serious-enough side effects, not just a minor side effect but a serious-enough adverse effect that resulted in the drug being withdrawn. I think that tells us that we're only just seeing the tip of the iceberg of this issue," Clayton tells me. This has become a huge concern for women's health activists, particularly in the United States, and has been one mandate of the Office of Research on Women's Health since 1990.

"As clinicians, we know very well that diseases show up differently in men and women. Every day, men and women present to the emergency room with different symptoms with the same condition," adds Clayton. "So heart attacks, for example, have different symptoms. Our research has shown that women who are going to have a myocardial infarction [heart attack] are more likely to have symptoms like insomnia, increasing fatigue, pain anywhere in the head all the way down to the chest, the weeks before they have a heart attack. Whereas men are less likely to have those symptoms and are more likely to present with the classic crushing chest pain." Given differences like these, she believes that excluding women from drug trials for so many years must have affected their health. "It's certainly a real possibility that the reason there are more adverse events in women than in men is because the whole process of drug discovery is tremendously biased towards the male," agrees Kathryn Sandberg, who researches sex differences in health, aging, and disease.

Again, though, this line of thinking risks drawing divisions between women and men when the picture of disease is far more complicated. While there's a clear benefit to better understanding women's bodies and having drugs that suit both sexes, the emphasis on sex difference starts to make it seem as though women's bodies are from Venus and men's bodies are from Mars. "Given the well-documented history of methodological problems with sex difference research, as well as harmful abuses of sex difference claims by those who would limit women's opportunities, it is remarkable to find women's health activists promoting, with little qualification, sex-based biology's expansive picture of sex differences," writes Harvard social scientist Sarah Richardson in her book *Sex Itself*.

But does it have to be one or the other? Is the only alternative to women being thought of as "little men" to have them treated as an entirely different kind of patient? As more detailed research is done, it's becoming clear that seeing some variation between women and men when it comes to health and survival doesn't mean we should ditch the notion that our bodies are in fact similar in many ways, too.

This is the cautionary tale of two drugs.

The first is digoxin, which has long been used to treat heart failure. In 2002 researchers at Yale University School of Medicine decided to take a look at the data around digoxin, analyzing its effects by sex. Between 1991 and 1996, researchers had carried out randomized trials on heart patients using digoxin. They found that it didn't affect how long a patient lived, but it did on average reduce their risk of hospitalization. But the Yale team noted that the drug was tested on roughly four times as many men as women, and they didn't respond identically. A slightly higher proportion of women taking digoxin died earlier than those taking a placebo. For men, the gaps between those taking the drug and the placebo group were much smaller. The sex difference, they concluded, "would have been subsumed by the effect of digoxin therapy among men."

But science never stands still. The Yale University result later turned out to be not what it seemed. More recent studies, including one with a much larger sample group published in the *British Medical Journal* in 2012, have suggested that in fact there isn't a substantially increased risk of death for women from digoxin use at all.

The second example is the insomnia drug zolpidem, commonly sold in the United States under the brand name Ambien. Sleeplessness is big business for pharmaceutical companies. Around sixty million sleeping pills were prescribed in the United States in 2011, up from forty-seven million just five years earlier, according to data collected by the health-care intelligence company IMS Health. And Ambien is among the most popular. Its side effects, however, include severe allergic reactions, memory loss, and the possibility of it becoming habit forming. The effects of zolpidem can also last longer than one night, leading to drowsiness the following day, which can in some cases make it dangerous to drive. Long after it was approved for market, research emerged that women given the same dose as men were more likely to suffer morning drowsiness. Eight hours after taking zolpidem, 15 percent of women but only 3 percent of men had enough of the drug in their system to raise their risk of a driving accident.

At the start of 2013 the US Food and Drug Administration took the landmark decision to lower the recommended starting dose of Ambien, halving it for women. "Zolpidem is kind of like a signal case," says Arthur Arnold.

However, just as with digoxin, the finding needed to be unpacked a little further. In 2014 research exploring the effects of zolpidem, carried out by scientists at Tufts University School of Medicine, suggested that its lingering effect may be mostly due to women's lower average body weight compared to men, which means the drug clears from their systems more slowly.

Digoxin and zolpidem highlight the pitfalls of including sex as a variable in medical research. Besides average body weight and height, women also have on average a higher percentage of body fat than men. And they generally take longer to pass food through their bowels. Both are things that might affect how drugs behave in their bodies. But they are also factors on which men and women overlap. Many women are heavier than the average man, for instance. It's not always the case that the sexes belong in two separate categories.

What also counts is the experience of being a woman, socially, culturally, and environmentally. "Both sex and gender are important factors for health," reminds Janine Clayton. Ideally, then, people should be treated according to the spectrum of factors that set them apart. Not just sex, but also social difference, culture, income, age, and others. As Sarah Richardson has written, "A female rat—not to mention a cell line—is not an embodied woman living in a richly textured social world."

The problem is that "medicine is very binary. Either you get the drug or you don't. Either you do this or you do that," says Sabine Oertelt-Prigione, who supports gender-based medicine. "So the only step, I believe, is to incorporate the notion that there is actually not one neutral body, but at least two. I believe it's just another way of looking at things. In medicine, just having a way to change paradigms and look at things differently can open up whole arrays of possibilities. It could be looking at sex differences, but there are many other things that could help to make health care more inclusive in the end."

"What are we trying to do? We're trying to improve human health, right?" adds Kathryn Sandberg. "So if we see a disease is more prevalent or more aggressive in men than women, or vice versa, we can learn a lot about that disease by studying why one sex is more susceptible while the other is more resilient. And this information can lead to new treatments that benefit all of us." Understanding why women tend to live longer could help men live longer. Including pregnant women in research may open up the cabinet of drugs that doctors can't currently prescribe because their effects on fetuses are uncertain. Medical dosages might be affected by a better understanding of how a woman's body responds across her menstrual cycle.

At the moment at least, the verdict of politicians and scientists seems to be that including sex as a variable when carrying out medical research *can* improve overall health. In 1993 the US Congress introduced the National Institutes of Health Revitalization Act, which includes a general requirement for all NIH-funded clinical studies to include women as test subjects, unless they have a good reason not to. By 2014, according to a report in *Nature* by Janine Clayton, just over half of clinical-research participants funded by the NIH were women.

Since the start of 2016 the law in the United States has been broadened to include females in vertebrate animal and tissue experiments. The

European Union now also requires the researchers it funds to consider gender as part of their work.

———————

For women's health campaigners and researchers like Janine Clayton and Sabine Oertelt-Prigione, this is a victory. To have females equally represented in research is something they've spent decades fighting for. Male bias, where it exists, is being swept away. Women are being taken into account. Maybe we will finally understand just what it is that makes women on average better survivors and why men seem to report less sickness.

But as science enters this new era, scientists need to be careful. Research into sex differences has an ugly and dangerous history. As the examples of digoxin and zolpidem prove, it's still prone to errors and overspeculation. As much as it can improve understanding, it also has the potential to damage the way we see women and drive the sexes farther apart. The research into genetic sex differences by people like Arthur Arnold doesn't just affect medicine but also how we see ourselves.

Once we start to assume that women have fundamentally different bodies from men, this quickly raises the question of how far the gaps stretch. Do sex chromosomes affect not just our health but all aspects of our bodies and minds, for example? If every cell is affected by sex, does that include brain cells? Do estrogen and progesterone not just prepare a woman for pregnancy and boost her immunity but also creep into her brain, affecting how she thinks and behaves? And does this mean that gender stereotypes, such as baby girls preferring dolls and the color pink, are in fact rooted in biology?

Before we know it we land on one of the most controversial questions in science: Are we born not just physically different but thinking differently too?

CHAPTER 3

A Difference at Birth

Girls and boys, in short, would play harmlessly together, if the distinction of sex was not inculcated long before nature makes any difference.

—Mary Wollstonecraft,
A Vindication of the Rights of Woman, 1792

"We live in jeans, don't we? They go with everything!" coos the mother. Her six-month-old daughter is wearing the tiniest pair of jeans I've ever seen, and she herself is dressed head to foot in denim.

We're sitting together in the baby lab at Birkbeck College in central London. It reminds me of a nursery, but a somewhat unusual one. A purple elephant decorates the door to a waiting area full of toys. Downstairs, meanwhile, a baby might be hooked up to an electroencephalograph that monitors her brain's electrical activity while she watches pictures on a screen. In another room, scientists could be watching a toddler play, examining which toys he happens to choose. Meanwhile, in this small laboratory that I've been invited into, a baby is being gently stroked along her back with a paintbrush. She's the thirtieth infant to be studied so far in this experiment.

"She really just likes sitting and watching, taking it all in. I'm happy sitting and observing, myself," her mother continues, bouncing the girl on her knee. Researchers suspect that human touch like this has an important impact on development in the early years. They just don't know how or why. So the goal of today's experiment is to measure how touch affects a baby's cognitive development. It's one of countless ways in which children are affected by their upbringing, slowly shaped into the people they will become.

Cute though babies are, studying them this way is not as much fun as it might seem. It's almost like working with animals. The challenge is to come up with clever experiments that get to the heart of their behavior without accidentally reading too much into what an infant does. A stare can be meaningful or mindless, while even the most charming a smile may just be wind. In this case, the researchers are using a paintbrush to run their touch experiment because that's the only way to control for parents stroking their children in different ways. With a brush, you can be sure it's the same every time.

Unfortunately, the baby's bottom lip begins to quiver and she erupts into tears. It's clear the paintbrush doesn't measure up to real touch. This is one result that can't be used.

"This is what baby science is. Trying to get a signal out of the noise," laughs Teodora Gliga, a psychologist at Birkbeck's Centre for Brain and Cognitive Development, who carries out research in the baby lab. Gliga's work focuses on how children develop in their early years, in the tradition of the Swiss psychologist Jean Piaget who, from the early twentieth century, observed his own children and famously realized that many of the assumptions scientists had made about early development were wrong. Babies aren't blank slates. Instead, he believed they are preprogrammed with their own ways of organizing knowledge about the world. The simplest example of this is a newborn's instinctive reflex to suck.

But this is just the start, scientists are realizing. The aim now is to figure out exactly how smart children are at birth, and what this means. One other use of baby research is to investigate differences between boys and girls. If children really are preprogrammed in some way, is the programming different depending on the sex? Do little girls prefer dolls dressed in pink because they're female or because society has taught them they should prefer dolls and the color pink?

Plenty of research has already been done. We know that around the age of two or three, children start to become aware of their own sex. Between the ages of four and six, a boy will realize that he will grow up to be a man and a girl that she will be a woman. It's also by then that children have some understanding of what's appropriate for each gender according to the culture they're in. American psychologist Diane Ruble and gender development expert Carol Lynn Martin have explained how, by the age of five, children already have in their heads a constellation of gender stereotypes. They describe one experiment in which children were shown pictures of people doing things like sewing and cooking. When a picture contradicted a traditional stereotype, the kids were more likely to remember it incorrectly. In one instance, instead of remembering that they had seen a picture of a girl sawing wood—which they had—some said instead that they'd seen one of a boy sawing wood.

Some parents are acutely aware of the problem. The mother of the baby I'm observing in the lab today tells me that she's a researcher with a PhD and she would like her daughter to have a PhD one day, too. Along the way, she's trying to avoid exposing her to gender stereotypes that might harm her sense of what she's able to do. "I'm not averse to pink, but we've tended to buy navy and blue things," she tells me. Someone offered to sell her a dolls' house recently, but she refused to take it. "I'd rather have something more neutral," she adds.

Researchers like those at Birkbeck College have realized that one of the most effective ways for scientists to sift nature from nurture, the biological from the social, is by studying children so young that they haven't yet been exposed to society's heavily gendered ways. "I don't think that studying adults tells us anything about sex differences. It tells us something about the lives those people lived. It's more about their experiences than about the biology of it," explains Teodora Gliga.

"The earlier you go in development, the closer you are to nature."

In 2000 a brief scientific article was published in the international journal *Infant Behavior and Development* describing an experiment that would shape the way people around the world thought about sex differences at birth. It was written by a team from the Departments of Experimental Psychology and Psychiatry at Cambridge University, which included Simon Baron-Cohen, a psychologist, neuroscientist, and famous expert on the medical condition autism. The paper claimed to prove for the first time that there were noticeable and important sex differences in the way newborn babies behaved.

The results were so powerful that they've been cited at least three hundred times in other research papers, as well as in books about pregnancy and childhood. When the then president of Harvard University, Lawrence Summers, controversially suggested in 2005 that the shortfall of female scientists and mathematicians might be because of innate biological differences between women and men, Simon Baron-Cohen used this study to defend him. Harvard University cognitive scientist Steven Pinker and London School of Economics philosopher Helena Cronin have both deployed it to argue that innate differences between the sexes exist. It has even made it into a Bible-inspired self-help book, *His Brain, Her Brain*, about how "divinely designed differences" between the sexes can help strengthen a marriage.

Since 2000, Baron-Cohen's department has made a formidable name for itself. At the time his paper was published, he was just two years away from unveiling a controversial and wide-ranging new theory about men and women, which he has named *empathizing-systemizing theory*. Its basic message is that the "female" brain is hardwired for empathy, while the "male" brain is built for analyzing and building systems, like cars and computers. People may show varying degrees of maleness and femaleness in their brains, but as the adjectives helpfully suggest, men on average tend to have "male" brains while women tend to have "female" ones.

Autism, which makes it difficult for people to understand and relate to others, is an extreme version of the male brain, adds Baron-Cohen. This is why people diagnosed with autism (until a few decades ago, they were almost all men, but many more women are now being identified with the condition, too) sometimes show unusual systemizing behavior, like the ability to do mathematical calculations in their heads very quickly or to memorize train timetables.

As yet, no one has been able to fully explain how, at the very start of a child's life, its brain gets set on a path toward being more male or more female. If such a mechanism is at work, the details are likely to be complicated. But according to Baron-Cohen, the crucial element is sex hormones—the chemicals at the root of many of the physical differences we see between women and men. Testosterone exposure in the womb, he argues, doesn't affect just the gonads and genitals but somehow also seeps into the male fetus's developing brain, molding it into a systemizing male brain. Female fetuses, which tend not to have as much testosterone, are left by default with empathizing female brains.

So then, what was the significance of his paper on newborn babies? Baron-Cohen wanted to see whether the stereotypes of women having stronger social skills and men being more mechanically minded might have a biological basis—in other words, whether girls are born empathizers and boys are born systemizers. For the first time anywhere, as far as he and his team were aware, they convinced the maternity ward of a local hospital to allow them to run a study on the youngest possible group of people. More than a hundred babies were included in the study, all around two days or younger, and all clearly far too young to be affected by social conditioning.

What they would observe, they claimed, would be nature untainted by nurture. And this made it a vitally important piece of evidence on which his empathizing-systemizing theory would hang.

Like many senior scientists do, Baron-Cohen left the experiment itself to a junior colleague, who had just joined his team. Jennifer Connellan was a twenty-two-year-old American postgraduate student. "I can't believe he accepted me into his lab actually," she tells me. By her own admission, she was young and inexperienced. Before arriving at Cambridge she was lifeguarding on a beach in California.

Each day, Connellan would turn up to the maternity ward to see if any mothers had given birth. The experiment itself was simple. "We wanted to contrast social versus mechanical," she says. So every baby was shown a face, which happened to be Connellan's own, and a mechanical mobile made from a picture of Connellan's face. They then measured how long every child looked at each one, if they looked at all. This long-established experimental method in baby research is known as "preferential looking." More socially inclined babies, the researchers hypothesized, would prefer to stare at the face, while more mechanically inclined babies might choose to look at the mobile. "It was quite rudimentary as far as the design," she recalls. "I felt like it was kind of like a science fair project."

When the results came in, a large proportion of babies showed no preference for the face or the mobile. But around 40 percent of the baby boys preferred to look at the mobile, compared to a quarter who preferred the face. Meanwhile, around 36 percent of the baby girls preferred the face, while only 17 percent preferred the mobile. It certainly wasn't the case that every boy was different from every girl, but, in research terms, the difference was statistically significant, enough for the scientific community to take notice.

In the published paper, Jennifer Connellan, Simon Baron-Cohen, and their colleagues argued that this was overwhelming evidence that boys are born with a stronger interest in mechanical objects, while girls tend to have naturally better social skills and more emotional sensitivity. "Here we demonstrate beyond reasonable doubt that these differences are, in part, biological in origin," they wrote.

"We were surprised that it was significant, that there was a significant difference," Connellan remembers. "[Baron-Cohen] was excited. I would say both of us were. We spent a lot of time going through it, making sure the results were what we thought they were." And sure enough, there it was, some of the seemingly strongest evidence yet that boys and girls really are born different. Cultural stereotypes about women being more empathic and men being more interested in building things might not just be due to the way their parents raised them and how society treated them.

"The fact that this was the earliest gender difference, that part was almost, like, shocking," she tells me.

_____

The next few years saw Simon Baron-Cohen put more meat on the bones of his idea that there are such things as distinctly female and male brains.

In 2003 he published *The Essential Difference*, a book written for the general public that lays bare what he sees as fundamental gaps between how men and women think. It includes a description of Connellan's experiment, along with pictures of her face and the mobile she showed the babies. "This sex difference in social interest was on the first day of life," he writes, adding elsewhere: "This difference at birth echoes a pattern we have seen right across the human lifespan. For example, on average, women engage in more 'consistent' social smiling." The clear implication is that the sexes don't appear to behave differently because of society or culture, but because of something profoundly innate and biological.

The differences, Baron-Cohen explains in his book, can be spotted in the types of hobbies people tend to choose.

> Those with the male brain tend to spend hours happily engaged in car or motorbike maintenance, small-plane piloting, sailing, bird- or train-spotting, mathematics, tweaking their sound systems, or busy with computer games and programming, DIY or photography. Those with the female brain tend to prefer to spend their time engaged in coffee mornings or having supper with friends, advising them on relationship problems, or caring for people or pets, or working for volunteer phone-lines listening to depressed, hurt, needy or even suicidal callers.

It's a slightly odd list. Peculiarly middle class and English, for one. It's also difficult not to notice that the male brain appears better suited to higher-paying, higher-status jobs like computer programming or mathematics, while the female brain seems to fit best with lower-status jobs, such as a caregiver or unpaid helpline worker.

Nonetheless, Baron-Cohen's ideas have been popular. His paper on the extreme male brain theory of autism has been cited more than a thousand times by other researchers. And the ideas behind empathizing-systemizing

theory have been widely mentioned by academics and intellectuals working in child development and gender. The eminent British biologist Lewis
Wolpert talks about his work in his own book on sex differences, *Why Can't
a Woman Be More Like a Man?*, published in 2014. "In general . . . the trend
may be summarised as males tending to think narrowly while females think
broadly," writes Wolpert.

Professor of biology and gender studies at Brown University Anne
Fausto-Sterling, however, is wary of research that claims to see sex differences in such young children. It's a controversial area of science, especially
given how unpredictable babies can be. It's also too easily swallowed by parents looking to understand their kids better, she adds. "You see it on baby
websites. You know, 'Expect your girl to do this, expect your boy to do this.'"
When scientists make these claims, argues Fausto-Sterling, they need to
be sure their findings are reliable. If Simon Baron-Cohen's work is taken
seriously, his ideas could have important consequences for the way society
makes judgments about what men and women should be doing with their
lives. "I think you end up having a theory that gives you permission to limit
both boys and girls to certain kinds of behaviors or longer term interests,
eventually vocations," she adds.

Simon Baron-Cohen was always aware that he was wading into divisive
territory. He writes near the start of *The Essential Difference* that he delayed
finishing it for years because he thought the topic was too politically sensitive. He makes the defense often made by scientists when they're publishing
work that might be interpreted as sexist—that science shouldn't shy away
from the truth, however uncomfortable it is. It's a claim that runs all the
way through work by people who claim to see sex differences. Objective
research, they say, is objective research.

"A lot of research findings never get replicated and are probably false."

When sex hormones were identified at the start of the twentieth century,
many scientists assumed they had just a fleeting effect on sexual behavior, the
same way we now realize that someone might get an adrenaline rush when
they're stressed or a surge of oxytocin when they're in love. As research
progressed, however, some began to suspect that there might be something
more permanent going on.

In 1980 two American researchers, psychologist and primate expert
Robert Goy and neuroscientist Bruce McEwen, published a survey of animal

experiments from preceding decades that explored the effects of testosterone levels around the time of birth. One study revealed that female rats given a single injection of testosterone on the day they were born showed less sexual behavior associated with females and more that associated with males when they became adults. Similar results were shown in rhesus monkeys, a species that's biologically not so far from humans and often used in research. A rhesus monkey was the first mammal sent into space, for instance. The more testosterone the monkeys were given, the more dramatic the differences.

Goy and McEwen's book *Sexual Differentiation of the Brain* claimed that testosterone has a lasting impact on future sexual behavior. But research like theirs couldn't be divorced from the age in which it was being done. Both science and gender studies had established the enormous role that culture plays in gender identity. In 1980 people commonly assumed that male and female brains were the same, and that behavioral differences in adults must be due to the way people were raised by their parents and shaped by society. One commentator compared talking about fetal testosterone and sex differences in the brain to talking about race and gaps in intelligence.

In an atmosphere like this, ideas like Goy and McEwen's marked a radical shift. And of course, they didn't go unchallenged. Critics pointed out, for instance, the bias in language being used to describe masculinity and femininity. Anything tomboyish, for instance, was interpreted as a girl behaving like a boy. But who was to say that this wasn't in fact a normal, common feature of being female? Others later complained that theories relying on primate studies for evidence didn't take into account that monkeys might treat their male and female offspring differently, as humans tend to do. If their genitals were affected by hormone treatment, this might affect how their mothers related to them, which might then have repercussions on their play or sexual behavior as adults.

Even though not everyone was comfortable with Goy and McEwen's findings, their line of research continued. It took its biggest leap with the controversial idea that the brain's entire structure might be shaped by testosterone levels in the womb, making men and women fundamentally different from birth—affecting not just sexual behavior but other behaviors as well.

Scottish neurologist Peter Behan and the US-based neurologists Norman Geschwind and Albert Galaburda said that studies on rats and rabbits showed how, even before a baby was born, higher than normal levels of testosterone slowed development on the left side of the brain, making the right side more dominant. Extended to humans, since boys naturally have more testosterone exposure before birth than girls do, it followed that men would be the ones who tend to have this larger right half of the brain. Interviewed by a reporter for the journal *Science* in 1983, Geschwind claimed that, if the mechanism connecting higher than usual levels of testosterone and the way a person responded to it was "just right, you get superior right hemisphere talents, such as artistic, musical or mathematical talent." It might explain, he implied, the higher numbers of world-class male rather than female composers and artists.

At the time, there was no medical way of safely measuring testosterone levels in a living fetus. So Geschwind instead relied on studying people who were left-handed (the right half of the brain tends to control muscles on the left side of the body, and vice versa, so someone with a dominant right half would be more likely to be left-handed). By this rough measure, at least one study at the time did indeed show slightly more left-handers among mathematically gifted children compared to the population in general.

In 1984 Geschwind and Galaburda published a book titled *Cerebral Dominance*, spelling out how their evidence supported the concept that men's brains were profoundly steered in a different direction by testosterone. And this is the very research that Simon Baron-Cohen has called upon in developing his own theory about empathizing female brains and systemizing male brains.

Geschwind died the same year that *Cerebral Dominance* came out. His death left the lingering question of whether he was right. Did the small amount of evidence in its favor mean that male brains really were hugely shaped by testosterone, or was the truth more complicated? "He was one of the most distinguished of neurologists," says Chris McManus, a professor of psychology at University College London, who has spent years dissecting the Geschwind-Behan-Galaburda theory. This was part of the problem with his work on testosterone and the brain, he adds. Geschwind's eminence in his field made it easy for his theory to be published in important journals, even when it turned out that the evidence for it was worryingly thin.

According to McManus, the Geschwind-Behan-Galaburda theory simply tried to do too much. At the time, it became a grand theory of how the brain was organized, drawing big connections between things that weren't necessarily connected, and between which the connections hadn't been proven. It was so broad that, even to this day, researchers have difficulty pinning it down. "If you're lucky, you can make it explain anything. . . . You can cut these things any way you want when you float free from data," says McManus.

But that doesn't mean that it was utter hokum.

Since the 1980s, detailed research using new techniques on animals does seem to suggest that sex hormones affect the brains of fetuses as they develop, leading to small differences in certain behaviors later on. It's a phenomenon that now has enough evidence behind it that neuroscientists and psychologists feel they cannot ignore it, even if this runs counter to their instincts. This is the unexpected nature of science: findings don't always sit happily with politics, and results are not always black and white. In this case, even though Geschwind's grand theory turns out to have been a little too grand, there may have been a kernel of a promising line of research hidden inside it.

In 2010 Cambridge University psychology professor Melissa Hines, who has carried out some of the world's most influential studies on sex and gender and is heavily referenced in Baron-Cohen's own papers, wrote in the journal *Trends in Cognitive Sciences* that thousands of experimental studies on nonhuman mammals show testosterone levels in the womb really do have an effect on behavior later on. Work like this is done by artificially injecting primates with extra hormones before monitoring their behavior. Her article includes a compelling pair of photographs, one of a female monkey inspecting a doll and the other of a male monkey moving a toy police car along the floor in a way that a child might.

But then, monkeys and humans are not the same. Making the leap from animals to people is critical to proving whether testosterone really does shape our own complicated minds in the same ways. If there *is* a similar difference, is it small, as it is in other mammals? Or is it large, in the way that Simon Baron-Cohen at Cambridge University suggests it is in his controversial empathizing-systemizing theory of male and female brains? Where does the truth lie?

Of course, the ethical standards for doing this research with humans are very different from primates. Scientists can't artificially inject a fetus or a child with more hormones to study the effects. Instead they must turn to people who have naturally very high or very low sex hormone levels. And these people are rare.

"I was unfinished when I was born," says Michael.

Michael isn't his real name, which I agree not to use. His real name isn't even his original name, which was Eilean. Michael's fifty-first birthday was two days ago, but he tells me he chooses not to celebrate it because he doesn't want to be reminded of the day he was born. That was the day his parents were told to raise him as a girl.

Michael was born a man, but a rare genetic condition meant that, at birth, his body didn't reflect this. Women typically have two sex chromosomes, known as "XX," and men have two called "XY." This Y chromosome is crucial because it helps prompt a fetus to produce androgens such as testosterone that make his body become obviously male inside the womb. Genes and hormones working together in this way are what make males look more male and females look more female. Michael is a regular XY male, but he has five-alpha-reductase deficiency, which means he's missing the enzyme that converts testosterone into a chemical that's crucial to developing the sex organs before birth. This means that he was genetically male, but his genitalia were ambiguous.

Cases like Michael's have helped biologists and psychologists get a grip on what it really means for humans to be born biologically one sex or another. If we want to know how sex hormones influence how masculine or feminine a person is, there's no better way than to study a person who is genetically male but whose body doesn't respond to hormones in the same way as the average man.

"When I was born, my sex wasn't determined at first look," he explains. "I had a penis but it was very, very small." It used to be common in cases like these for doctors to advise people like Michael to live their lives out as girls, because surgery to make their genitals appear female is simpler than constructing a penis. When Michael was born, experts believed that gender was so heavily shaped by society that this was a perfectly reasonable choice to make. If he were treated like a girl, he might feel like one. Some children in similar situations have adapted to their new gender identities. But for many, including Michael, decisions like these have led to personal tragedy.

His underdeveloped testes were left inside his body, before being partly removed when was five years old, long before puberty set in. This surgery was accidentally left incomplete, which meant that he was still producing small amounts of testosterone. The whole time, he was oblivious to his genetic sex. To the world he was a girl, but inside he became increasingly aware that he didn't feel like one.

At around the age of three, he started showing an interest in typical boys' toys. Later in his school's physical educational lessons, when girls were told to go to one side of the sports hall and boys to the other, he would stand in the middle, uncertain. "The teachers kept separating me off from the boys," he remembers. For a young boy, the situation was as tragic as it was confusing. Another time, when a shopkeeper asked him, "What can I get for you, son?" he imagined in delight that she must have seen him for who he truly was. When someone behind him explained that he was actually a girl, it felt like a slap in the face. "As I got older, I looked at my grandmother, and mother, and female cousins and realized I will never be like them," he recalls.

His childhood was an impossible confusion, trapped between what society expected of him—including being repeatedly told, "Girls don't behave like that!"—and his personal conviction that he was a boy. He remembers his shame when, as a member of a choir as a child, his voice began to break and he had no choice but to blame it on a sore throat. When he was much older, people assumed he was just a very athletic girl. "People identified me as a tomboy," he explains.

People with conditions like Michael's are today described as "intersex." It's an umbrella under which many extremely rare conditions sit, including androgen insensitivity syndrome, in which a person with male chromosomes appears entirely female because their body doesn't recognize testosterone, and congenital adrenal hyperplasia, in which women are born looking female but have high levels of male hormones, which can cause ambiguous genitalia. They're not eunuchs or hermaphrodites. They don't fit the binary categories of male and female, but instead occupy a biological middle point, which many people have yet to accept or understand.

"I have seen less than ten cases in my entire career," says British endocrinologist Richard Quinton of androgen insensitivity syndrome, one condition he treats. A career spent observing people with intersex conditions, along with others who want to change gender, has given Quinton a special insight into how hormones affect sexual identity. Many patients choose to keep quiet about their conditions. But Quinton heard of an instance once in the Middle East where two sisters, both with androgen insensitivity syndrome, brought a case before the Islamic courts to be recognized as men so they could secure their family inheritance, which wouldn't be passed down to them if they were women. With congenital adrenal hyperplasia, he says, "at the extreme end you can have some births that tend to look male," although most look female with some masculine features. These patients "are said to be more tomboyish in their behavior, certainly in childhood. And when older, many are also attracted to the same sex."

At sixteen years old, after finding out his true medical history, Michael finally had a chance to make his own decision about how to live the rest of his life. At nineteen he began transitioning into a man, taking weekly testosterone injections. His voice got deeper; hair grew on his arms, legs, and face; and he developed more muscle. "It was like the sun coming out," he says.

The genital surgery inflicted on him when he was born was described at the time as "tidying up," but he sees it now as child abuse. "What happens with a lot of these children is that they grow up in confusion," says Michael, who has since found acceptance and understanding through the support group UK Intersex Association.

Today Michael is a psychologist with a specialty in child mental health, a career he chose partly because of his own experiences. His voice is strong and clear. His gender is indisputably male. He is also living evidence that at least some aspects of gender identity must be rooted in biology. Hormones don't just affect how our bodies look, but how we perceive ourselves, too. The question this then raises is how much more of an effect do hormones have on how we think and behave? How much do testosterone, estrogen, and progesterone shape our minds and steer them in different directions?

———

I'm told that psychology professor Melissa Hines is one of the most balanced and fair researchers in her field—which is important in a field that is sometimes neither balanced nor fair. Her office, at the end of a warren of old, wooden corridors behind a small lane in Cambridge, is lined with books about gender from all sides of the debate. She chooses her words carefully.

"We've looked at a variety of behaviors," she begins. Hines relies on intersex cases like Michael's to carry out her research on the effects that hormones have on psychological sex difference, including intelligence. Like baby research, this is an important part of the equation when it comes to understanding nature and nurture. If testosterone does steer boys toward having a distinctly male brain, different from a female brain, then we should see clear differences in how people with unusually high or low testosterone behave.

Her findings reveal three areas that show a statistically marked difference. Starting with the obvious first, "for gender identity, the differences are huge. Most men think of themselves as men and most women don't," she states. "The second thing is sexual orientation. Most women are interested in men, and most men aren't." The third one is childhood play behavior. Studying girls with congenital adrenal hyperplasia, with higher than normal levels of testosterone, she found, "Rough-and-tumble play is increased in girls exposed to androgens. They like boys' toys a bit more, girls' toys a bit less, and they like to play with boys more than the average girl does, but not as much as the average boy. That's been replicated by seven or eight independent research teams."

The fact that research is replicated is crucial. A lot of work in the field of psychology, even the most widely reported on in the press, hasn't been. If a number of independent scientists come to the same conclusions based on different studies across a broad range of people, then it's far easier to be confident about the results. "A lot of research findings never get replicated and probably are false," she admits. "It's just the way science works. You can't study the whole world, so you have to take a sample, and your sample may or may not be representative." This is so important to Hines that, when I meet her, she goes so far as to warn me that she isn't even sure about the reliability of some of her own research because it hasn't yet been replicated elsewhere.

On toy preferences, now, she has little doubt left. "One of the first studies I did in this area was bringing children into the playroom with all the toys and just recording how much time they spend playing with each toy," she describes. "I was really surprised by the results because, at the time, the thought was that toy choices are completely socially determined. And you can see why, because there is so much social pressure for children to play with the gender-appropriate toy." She and others found in study after study that boys on average really do prefer to play with trucks and cars, while girls on average prefer dolls. "The main toys are vehicles and dolls. Those are the most gendered type of toys," she says.

A study that Hines and her colleagues carried out on infants in 2010, watching for how long children look at one toy over another, suggested that these preferences might start to emerge close to the age of two. "Between twelve and twenty-four months, children were already showing preferences for sex-typed toys. So, the girls were looking longer at the dolls than at the car, and the boys were looking longer at the car than at the doll," she says. But at twelve months, both boys and girls spent longer looking at the doll than the car.

Statistically, this difference in how young children play is significant. "Toy preferences, I like to compare to height," she explains. "We know that men are taller than women but not all men are taller than all women. So the size of that sex difference is two standard deviations. The sex difference in time playing with dolls versus trucks is about the same as the sex difference in height." A standard deviation is a measure of how spread out data are.

The spread of height looks like a bell-shaped curve. The average height of men is around sixty-nine inches and the standard deviation is three inches. This means that, in a large group of men, more than two-thirds will be within one standard deviation of the average, making them between sixty-six and seventy-two inches tall. The farther you get from the average, toward the thin ends of the bell curve, the fewer men there are. Two standard deviations away will be men who are six inches taller or six inches shorter than the average man (less than 5 percent of men are two or more standard deviations away from the average). A difference in behavior of two standard deviations between men and women would therefore be like a difference of six inches between their average heights. In everyday life, it's a noticeable gap.

In studying girls with congenital adrenal hyperplasia, Hines's team was keen to test whether they might be getting some unconscious encouragement to play with boys' toys, perhaps because their families knew of their intersex condition. "So we thought, let's bring parents in with them and see how they react. Are they encouraging the girls to play in that way or not in the playroom?" she says. "But we found what they actually did was try to get them to play with female typical toys. More so than with their other daughters, they would introduce female toys. If she was playing with a female toy they would say, 'That's nice,' and give them a hug." It's more evidence, she implies, that the differences they've seen in toy preferences aren't purely due to social conditioning but have a biological element, too.

This difference in toy choices, however, is a far leap from the theory that the brains of men and women are deeply structurally different because of how much testosterone they've been exposed to. It's also a considerable distance from Baron-Cohen's claim that there's such a thing as a typical male brain and a typical female brain—one that prefers mathematics and another that likes coffee mornings. For him to be right, there would have to be noticeable gaps in lots of other behaviors as well. Those with female brains would have to clearly behave on average like empathizers and those with male brains like systemizers.

According to Hines, this isn't what we see. Tallying all the scientific data she has seen across all ages, Hines believes that the "sex difference in empathizing and systemizing is about half a standard deviation." This would be equivalent to a gap of about an inch between the average heights of men and women. It's small. "That's typical," she adds. "Most sex differences are in that range, And for a lot of things, we don't show any sex differences."

Researchers have known this for a long time. In their 1974 book *The Psychology of Sex Differences*, American researchers Eleanor Maccoby and Carol Nagy Jacklin picked through an enormous mass of studies looking at similarities and differences between boys and girls. They concluded that the psychological gaps between women and men were far smaller than the differences that existed in society among women and among men. In 2010, Hines repeated this exercise using more recent research. She found that only the tiniest gaps, if any, existed between boys' and girls' fine motor skills, ability to perform mental rotations, spatial visualization, mathematics ability, verbal fluency, and vocabulary. On all these measures, boys and girls performed almost the same.

Teodora Gliga from the Birkbeck baby lab agrees that when it comes to children raised under normal conditions, without unusual medical conditions, large gaps between girls and boys haven't been found. "It's quite rare to find differences in typical development." The overlap between the sexes is so huge, she explains, that scientists have struggled to find and replicate results that suggest that there is a real gap between the sexes. "For the time being, the baby science is not convincingly showing any consistent differences."

Even studying the tiny minority of girls who have been exposed to higher than usual levels of androgens, adds Hines, while it does tell us something about sex differences, doesn't tell us that these differences are particularly big. "If genetically I am a girl fetus that produces a bit more androgen, maybe I'll play a bit more with boys than if I had a bit less. Then maybe I'll have two friends who are boys, instead of one." Beyond gender identity and toy preference, on pretty much every other behavioral and cognitive measure that scientists have investigated (in a field that has left few stones unturned), girls and boys overlap hugely. Indeed, almost entirely. In a study by Hines exploring color preferences, for example, she found infant girls also had no more of a love of pink than boys did.

In 2005 University of Wisconsin, Madison, psychologist Janet Shibley Hyde proposed a "gender similarities hypothesis" to demonstrate just how big this overlap is. In a table more than three pages long, she lists the statistical gaps that have been found between the sexes on all kinds of measures, from vocabulary and anxiety about mathematics to aggression and self-esteem. In every case, except for throwing distance and vertical jumping, females are less than one standard deviation apart from males. On many measures, they are less than a tenth of a standard deviation apart, which is indistinguishable in everyday life.

When it comes to intelligence, too, it has been convincingly established that there are no differences between the average woman and man. Psychologist Roberto Colom at the Autonomous University of Madrid, Spain, found negligible differences in "general intelligence" (a measure that takes into account intelligence, cognitive ability, and mental ability) when he tested more than ten thousand adults who were applying to a private university between 1989 and 1995. His paper, published in the journal *Intelligence* in 2000, confirms what earlier studies have repeatedly shown.

Some have argued that there is statistically more variation among men than among women, which means that even though the average man is no more intelligent than the average woman, there are more men of extremely low intelligence and more men of extremely high intelligence. At the far ends of the bell curve where the overlap ends, they say, the difference becomes clear. This may have been the basis for the controversial point made by Harvard president Lawrence Summers in 2005 when he was hunting for explanations for why there are so many more male than female science professors at top universities.

Studies haven't fully supported this explanation. In 2008, using population-wide surveys of general intelligence among eleven-year-olds in Scotland, a team of researchers based at the University of Edinburgh confirmed that males did show more variability in their test results. These differences aren't extreme as some in the past have suggested they are, they note, but they are substantial. At the same time, the authors point out that the biggest effect is seen at the bottom end of the scale. Those with the very lowest intelligence scores tend to be male. This is partly genetic. X-linked mental retardation, for instance, affects far more men than women.

"Mainly it's at the bottom extreme because they have more developmental disorders," explains Melissa Hines. "At the upper extreme, it's not that big a difference." The authors of the Scottish study showed that the smaller differences they saw at the top end certainly weren't enough to account for the gaps between women and men taking up mathematics and science. In their particular set of data, around two boys for every girl achieved the very highest intelligence test scores. At universities, gaps in the numbers of male and female science professors are usually far bigger.

Hines adds that this difference in Scottish test results could also be due to social factors. "Even though on the average there is no sex difference in IQ, I think still boys get encouraged at the top. I think in some social environments, they don't get encouraged at all, but I think in affluent, educated social environments, there is still a tendency to expect more from boys, to invest more in boys," she tells me.

This observation is backed up by recent research into how people often think of genius as being a male feature. A 2015 study published in the journal *Science* explored whether this expectation of raw brilliance in men might affect the gender balance in certain subjects. Led by the Princeton University philosophy professor Sarah-Jane Leslie and University of Illinois psychologist Andrei Cimpian, the researchers asked academics from thirty disciplines across the United States if they believed being a top scholar in their field required "a special aptitude that just can't be taught." They found that in those disciplines in which people thought you did need to have an innate gift or talent to succeed, there were fewer female PhDs.

The subjects that instead valued hard work tended to have more women.

"It's hard to separate our opinion from the data."

Perhaps naive, Jennifer Connellan didn't expect the backlash when it came. But then, no one could have expected that, when it came, it would be so huge.

Not long after her and Simon Baron-Cohen's study on newborns preferring faces or mobiles was published in 2000, people began to question their research. Could it be true that there was such a deep sex difference in the behavior of newborn babies? Were girls really preprogrammed to be empathizers while boys were born systemizers? Flickers of doubt were raised about her methods and the reliability of the results.

The skepticism came to a head in 2007 when New York psychologists Alison Nash and Giordana Grossi dissected the experiment in forensic detail and catalogued a string of problems, big and small. For one thing, the paper's grand claim that the experiment's conclusions were "beyond reasonable doubt" seemed an uncomfortable stretch when, in fact, not even half the boys in the study preferred to stare at the mobile and an even smaller percentage of the girls preferred to stare at the face.

But their most damning criticism was that Connellan knew the sex of at least some of the babies she was testing. This could have caused any number of subtle biases. For instance, consciously or not, she may have moved her head to make the girls look at her longer, Nash and Grossi pointed out. The need to avoid this sort of problem is exactly why scientists are advised to carry out these studies blind, without knowing the sex of their subjects. Without this safety measure, it's hard to take the results seriously.

Psychologist and author Cordelia Fine, who in 2010 published *Delusions of Gender*, a book about the problems with brain research that includes Nash and Grossi's findings, adds that, even if their findings were right, Connellan, Baron-Cohen, and their colleagues made too big a leap when speculating about what they might mean. "One assumption is that these visual preferences predict a child's later empathizing versus systemizing interests, for which there is no evidence either way," she tells me.

When I put these criticisms to Connellan herself, now fifteen years since her paper was published, she accepts them humbly. At the time, her paper was out before she had been awarded her doctorate, and the flood of criticism came to bite when she turned up to defend her work in front of a panel of reviewers. She was told she had failed. "To have the defense go as poorly as it did was really surprising," she says. She attributed it to "lots of politics in there with the reviewers. . . . We appealed it and got some more neutral people." Only then, with a new set of reviewers, did she finally pass.

The experiment did have its problems, she admits. She found it impossible to prevent herself from being aware of the sex of some babies, mostly because she was in a maternity ward surrounded by newborn paraphernalia, including pink and blue balloons, and sometimes even their names. "We were testing the babies in a neutral zone, where there were no balloons or anything like that, and the blankets were all neutral. That was actually where we did the experimentation," she says. But in getting permission to test the babies, they had to go see the mothers first, in an environment that was far from neutral.

"We did the best we could with the results that we had," she admits. "Are they perfect? No." In writing the paper, too, she says that she may have become overexcited by the results. "I was very inexperienced, and I think that inexperience caused more of the problems than anything else."

When I ask Simon Baron-Cohen to give me his own thoughts on the experiment, he tells me by e-mail, "It was designed thoroughly and was scrutinised through peer review and as such it met the bar for good science. No study is above criticism in the sense that one can always think of ways to improve the study, and I hope when a replication is attempted, it will also be improved."

In fact, replication has been one of the biggest problems for the experiment. To date, nobody has attempted to copy it to check if the findings were reliable. "Studies have to be replicated," comments Teodora Gliga, "especially if it's a new idea. It needs to be replicated, otherwise it's not believable.

It's an interesting idea, but not a fact." Subsequent studies with slightly older children have shown no sex differences. And, as Melissa Hines's work has revealed, there appear to be no toy preferences among children until at least the age of one, and possibly closer to two years old.

Baron-Cohen, however, tells me that "the fact that the study hasn't yet been replicated does not invalidate it at all. It simply means we are still awaiting replication." One explanation he gives for why no other researchers have tried to copy it is that babies are difficult to test, which means you need large groups to get a reliable result. "Second, it appears that testing for psychological sex differences in neonates still attracts a fair amount of controversy. So some researchers may have been deterred by not wanting to walk into a potential political minefield," he adds.

Jennifer Connellan has since abandoned the minefield altogether. Her career in Simon Baron-Cohen's lab turned out to be brief. After getting her degree, she left Cambridge to join Pepperdine University. Today, she runs a tutoring company in California. She's also mother to a girl and a boy. She tells me that she remains intrigued by the idea of empathizing and systemizing brain types, but believes that it's only at the extremes where researchers seem to find any discrepancies. "It's all a bell curve . . . and for the kids in the middle there's almost no sex difference there at all," she says.

Baron-Cohen, meanwhile, presses on in trying to establish links between levels of testosterone before birth and sex differences in the brain. In 2002 he and another postgraduate student, Svetlana Lutchmaya, claimed that twelve-month-old girls they observed in experiments made more eye contact than boys of the same age did. This study has been cited by other researchers more than two hundred times.

Then in 2014 Baron-Cohen and his colleagues published the results of a study looking at one of the biggest sources of data in the world: more than nineteen thousand amniotic fluid samples in Denmark, taken from pregnant women for medical reasons between 1993 and 1999. If ever a set of data could reliably prove his hypothesis that high fetal testosterone levels are linked to autism, leading to the "extreme male brain," it was this one. His team managed to measure hormone levels in these fluid samples to find out how much testosterone the babies would have been exposed to. They could then crosscheck all this with the medical and psychiatric records of the same set of children when they were older. It was an amazingly large and thorough set of patient information.

The database included 128 males who were diagnosed with a condition on the autism spectrum. But Melissa Hines tells me that Baron-Cohen's results didn't show a direct link between them and high fetal testosterone levels. "That was like the ultimate test, and there was no correlation between testosterone and getting an autism spectrum diagnosis," she says. "That's just one study, but it doesn't support it."

Without evidence of a clear connection between the "extreme male brain" and testosterone, when their findings were published in the journal *Molecular Psychiatry* in 2014, Baron-Cohen and his colleagues instead claimed to see a correlation between autism and a mixture of hormones, including testosterone, but also the female sex hormones, progesterone and estrogen. He tells me the reason they did this is because "the sex steroid hormones in that pathway are not independent of each other because each is synthesized from its precursor so that the level of one hormone will directly affect the level of the next one in the pathway."

Hines has since run her own study of correlations between fetal testosterone levels and autistic traits on children with congenital adrenal hyperplasia, which was published in the *Journal of Child Psychology and Psychiatry* in 2016. She found no link.

I can't help wondering what Hines thinks is going on in her own field. She falls short of using the word *sexism*, but she does believe scientists haven't always done as good a job on sex and gender differences as they could have done. "I don't think people do it intentionally. I think these are things we deal with every day," she says. Gender is one of those subjects that everyone has an opinion on, and of course, of which everyone has direct experience. Perhaps unsurprisingly then, there's sometimes a lack of objectivity in the field.

"It's hard to separate our opinion from the data," she warns. "I think this is something the human mind does. It wants to have things that define maleness and things that define femaleness. Now maleness, historically in psychology, has been instrumentality, so that's kind of like systemizing, and femaleness has been nurturing, warmth, kind of like empathy. So there is a long tradition of conceptualizing this in similar ways. . . . But I'm not sure where it gets us, because there's lots of overlap. So you can't give someone a test and get these scores and say they're male or female. There's too much individual variability."

"I think we really have to be extraordinarily careful . . . when we talk about overlapping populations with huge variability," agrees Brown University's Anne Fausto-Sterling, one of the world's leading researchers on gender.

She believes that Simon Baron-Cohen's theory of male and female brains makes little sense. Connecting testosterone levels before birth to behavioral sex differences later on, she says, "is just this huge explanatory leap, and it leaves me uncomfortable because I don't think it's much of a scientific explanation when you make such a big leap. . . . We do see the differences, and I don't disagree with that finding. What I disagree with is leaping to the idea that that this means it is something innate or inborn," she adds. "I do think that if you just jump to the prenatal . . . you miss a whole developmental window when something very important and very social is going on."

Fausto-Sterling belongs to a vanguard of biologists and psychologists who see the nature versus nurture question as old-fashioned. "There is a better way of looking at the body and how it works in the world, and understanding the body as a socially formed entity, which it is," she explains. Men and women may be different, but only in the same way that every individual is from the next. Or, as she has also put it, "that gender differences fall on a continuum, not into two separate buckets."

"I think that people tend to think of this in an either-or kind of way," agrees Teodora Gliga. Either girls and boys are born very different or they're the same. The scientific picture emerging now is that there may be very small biological differences, but that these can be so easily reinforced by society that they appear much bigger as a child grows.

"My opinion is that you will find differences wherever they were reinforced, because we love categories, . . . we need to have categories. And so once we've decided, once we've labeled 'this is a girl,' 'this is a boy,' then we have so many culturally strong biases that we maybe produce differences in abilities. So for example, in physical abilities, if we push boys to be more active and to deal with danger, then of course later in life when they're children, they will look different. But that does not mean the differences were in the biology," says Gliga.

Instead of the binary categories we have now, Fausto-Sterling believes that every individual should be thought of as a developmental *system*—a unique and ever-changing product of upbringing, culture, history, and experience, as well as biology. Only this way, she argues, can we truly get to the heart of why women and men across the world appear to be so different from each other, when studies into mathematics ability, intelligence, motor skills, and almost every other measure consistently tell us they're not.

If toy preferences don't emerge until at least age one and other differences reveal themselves even later, she suggests, then what else could be happening up until age one? One line of research that hasn't been fully explored, for example, is counting exactly how many toys babies are given in the first year of life, and what kinds of toys they are. "I can say that boys see more boys' toys and girl see more girls' toys, but honestly there is no data to show that," she says.

In her most recent research project, Fausto-Sterling has tried to get closer to answers by filming mothers playing with their children. She recounts one vivid example: "You see a little three-month-old boy, just slouched on the couch. He's not even big enough to sit up on his own, but he's kind of propped up with pillows. His mother is trying to engage him in play, and she's stuffing little soft footballs in his face, American footballs. . . . She's thrusting this football at him and saying, 'Don't you want to hold the football? Don't you want to play football like your daddy does?' And he's just sitting there like a kind of blob. He has no interest one way or the other," she describes.

The impact of actions like these, small as they may seem, can be long lasting. "If that kind of interaction is going on iteratively in the early months, then if at some point he does reach out and grab, when he's big enough to do that, at four months, five months, or six months, he's going to get a very positive reinforcing response from his mother," Fausto-Sterling explains. This relationship between the boy and footballs is strengthened as he sees how happy they make his mother, and also because the toy is already so familiar to him. "He may see them again at an older age, when he is more capable of physically interacting with them. And just seeing them and recognizing them may give him a certain kind of pleasure." By the end, the boy appears to love football.

Fausto-Sterling adds that evidence is emerging from her team's observations of mothers that boys are also handled differently from girls, which might be influencing the way they grow. "The mothers of sons in my cohort are moving them around a lot more. They're shifting them, they're playing with them, and they're talking to them less. They're more affectionate to them when they're moving them physically." This could simply be because boys demand more physical movement from the start, but again, it's another element of the development process that hasn't been fully studied.

Work like hers, while in its early days, reinforces that countless little thumb marks are in the ball of dough that is a developing child. Hormonal effects on the brain or other deep-seated biological gaps aren't necessarily the most powerful reason for the gaps we see between the sexes. Culture and upbringing could better explain why boys and girls grow up to seem different from each other.

And if this is the case, a change in culture or tweaks to upbringing might reverse the differences. "If you see what you think is a disability, don't understand how it developed in the body and where it came from. Understand that bodies are shaped by culture from the very get-go," explains Fausto-Sterling. "If you neglect a child at birth, their brain stops developing and they're pretty messed up. If you highly stimulate a child, if they're within a normal developmental range, they now develop all sorts of capacities you didn't know they had or didn't have the potential to develop. So the question always goes back to how development works."

Melissa Hines agrees that there's no reason nature should determine a girl's destiny, despite her studies showing that testosterone may explain some small behavioral sex differences. "I do believe that testosterone prenatally sets things in motion in a certain direction, but that doesn't mean it's inevitable. It's like a river. You can change its course if you want to," she tells me.

———

Changing the river's course is easier than it seems. It depends on society wanting to change in the first place. And this is a world in which even cold, rational scientists can't abandon their desire to hunt for differences between women and men. The effects of testosterone on the brain are just one example. In 2013 a team from Taiwan, Cyprus, and the United Kingdom (in which, incidentally, one member was neuroscientist Simon Baron-Cohen) highlighted another. They got together a large number of independent studies into sex differences in brain volume and density to see what they could tell us in summary. In their paper published the following year, the team proclaimed that men's brains were typically bigger by volume than women's brains. The gap ranges from 8 to 13 percent.

This isn't news. It's long been known that men have on average slightly bigger heads and slightly bigger brains than women. It's a finding that's been popping up in scientific journals for more than a century.

But it indicates a problem that doesn't go away, no matter how much time passes. Brain researchers have never been able to resist the urge to scour the skulls of women and men in search of variation. And the reason they persist with this endeavor is simple. Because if a man's brain looks physically different from a woman's, well then, perhaps this will confirm that something different is going on in their minds, too.

CHAPTER 4

The Missing Five Ounces of the Female Brain

*The clearness and strength of the brain of the woman prove continu-
ally the injustice of the clamorous contempt long poured upon what
was scornfully called "the female mind."*

—Charlotte Perkins Gilman, *Women and Economics,* 1898

On the twenty-ninth of September, 1927, a dead brain made the news. It
appeared on page five of the *Cornell Daily Sun.*

Before I tell you why, let me tell you about this brain's owner. It belonged to the teacher and writer Alice Chenoweth Day, who by the time of
her death was better known by the pen name Helen Hamilton Gardener.
Since 1875 Gardener had lived with her husband in New York, where she
was a passionate advocate for the rights of women. One of her books, *Facts
and Fictions of Life*, railed against the way women were kept subservient by
society though unequal education and marriage.

Gardener's work echoed that of suffragist and writer Eliza Burt Gamble,
who was her contemporary. She, too, was incensed by the way scientific
"facts" were being used to hold women back in their fight for equality. In
1888 Gardener gave a talk titled "Sex in Brain" at the convention of the
International Council of Women in Washington, DC, complaining that scientists studying the brain claimed that women's brains were lighter than
men's, and that by extension, they must also be less intelligent. One of the
most high-profile men to suggest this was William Alexander Hammond,
no less than surgeon general in the US Army and one of the founders of the
American Neurological Association.

Gardener didn't have the education she needed to prove that Hammond was wrong. Few people, she lamented, "had the anatomical and anthropological information to risk a fight on a field which assumed to be held by those who based all of their arguments upon scientific facts, collected by microscope and scales and reduced to unanswerable statistics." If scientists wanted to make such outrageous assertions, what could she or any other layperson do to fight them?

"I finally, with fear and trembling, made up my mind to learn what he knew on this subject or perish in the attempt," she announced. She ended up working alongside New York doctor Edward Spitzka, soon to become president of the American Neurological Association, in the hope of understanding the brain's anatomy enough to be able to challenge the great William Hammond. It took her fourteen months to dissect his statistics, while corresponding with twenty anatomists and doctors across New York.

In a beautifully clever and witty letter eventually published in *Popular Science Monthly*, she revealed that all her experts couldn't distinguish between a male and female brain at birth. Even among adults, it would be a mere guess whether a given brain was male or female. The overlap between the sexes was just too big. Her sharpest observation was that the weight of a person's brain couldn't be a measure of intelligence, anyway. It was the ratio of body weight to brain weight or body size to brain size that was important. If that weren't the case, she remarked, "an elephant might out-think any of us." Indeed we should expect a creature as huge as a whale, with its correspondingly huge brain, to be a genius.

Her arguments were compelling, but apparently not compelling enough. William Hammond replied to Gardener with a hefty five-page letter of his own (he complained that he nearly didn't write it at all because he found the tone of hers was "so bad"). Mocking her "twenty leading brain anatomists," he repeated his own results. He added, "Ten men who were remarkable for their intellectual development" were found to have particularly heavy brains, on average weighing more than fifty-four ounces. "Now, let Miss Gardener and the 'twenty leading brain-anatomists,' etc., search the records of anthropology and their own immense collections for the brain of a woman weighing as much as the least of these," he challenged.

A month after her letter was published, George John Romanes, an eminent evolutionary biologist and friend of Charles Darwin, also weighed in. "Seeing that the average brain-weight of women is about five ounces less than that of men, on merely anatomical grounds we should be prepared to expect a marked inferiority of intellectual power in the former," he argued in *Popular Science Monthly*. "We must look the facts in the face. How long it may take the woman of the future to recover the ground which has been lost in the psychological race by the woman of the past, it is impossible to say; but we may predict with confidence that, even under the most favourable conditions as to culture, and even supposing the mind of man to remain stationary, . . . it must take many centuries for heredity to produce the missing five ounces of the female brain."

The fight over those missing five ounces was a bitter one, and it was never resolved in Helen Hamilton Gardener's lifetime. Scientists like William Hammond and George John Romanes "gave a black eye to their facts in preserving a blind eye to their faith," she warned.

Gardener promised, fittingly, to leave her brain to science before she died. In 1925 it ended up in the Wilder Brain Collection at Cornell University (it's still there, preserved in a jar). And this is how the *Cornell Daily Sun* happened to feature an article about Helen Hamilton Gardener in 1927. When it was studied, her brain weighed in at 1,150 grams (approx. 2.5 pounds), around five ounces less than the average male brain. But this didn't mean she wasn't vindicated. "In the structure of her own brain Mrs. Gardener has presented abundant evidence that the brain of a woman need not be inferior to that of a man of equal rank," the newspaper report proclaimed. Hers happened to weigh the same as that of esteemed anatomy and neurology professor at Cornell, Burt Green Wilder—the very founder of the brain collection itself.

Gardener's point was made. Today it's well established that brain size is related to body size. Paul Matthews, the head of brain sciences at Imperial College London, tells me, "If you correct for skull size, there are very tiny differences between the two sexes, but their brains are much more similar than they are different." The missing five ounces are accounted for.

But that hasn't stopped scientists, even today, combing brains for evidence that women think differently from men.

"Males will have an easier time seeing and doing."

"When did you first become interested in studying sex differences?" I ask Ruben Gur, professor of psychology at the Perelman School of Medicine at the University of Pennsylvania. He pauses. "Since adolescence! Before that, I wasn't that interested," he jokes.

Ruben is one of two Gurs, the other being his collaborator and wife, Raquel Gur (a professor of psychiatry in the same school, who doesn't respond to my request for an interview), who have dedicated their careers to understanding how the brains of women and men differ and what this means. Their first experiment in this niche was published in 1982, when Ruben Gur was thirty-five years old. Measuring blood flow through the brains of healthy people they found, to his surprise, that women had 15 to 20 percent higher flow rates than men. It was such an unexpected result, he tells me, that CNN was outside his lab the next morning for an interview.

This marked the start of a long string of headline-grabbing scientific publications. And their timing was perfect. In the 1970s sex difference research had experienced a decline because gender scholars and women's rights campaigners argued that it was sexist to look for biological gaps between women and men, just as it was racist to look for differences between black and white people. Gradually, though, it became acceptable again. Neuroscience is a field in its infancy when judged by the task it has ahead of it. The brain is as dense and complex a thing as anyone has ever studied, with billions of nerve cells and an impossibly sophisticated web of connections between them. But understanding has recently been improved thanks to new imaging technologies, which allow scientists to understand brain activity in more detail than ever. These technologies have reinvigorated the search for difference. In 2006 the Gurs were invited to appear on the *Today* show to use one of these scanners to spot differences between the show's medical editor's brain and that of her husband.

Looking for sex differences in the brain isn't just socially acceptable nowadays, it's almost fashionable. "Back in 1982, we were lone wolves in the wilderness. Now everybody's doing it!" laughs Ruben Gur.

What has changed since the nineteenth century, though, isn't only the technology but also what we know about what's inside our skulls. Researchers can no longer weigh or measure brains like lumps of coal, then assume this tells them something about human behavior or intellectual capability. "Of course the male brain looks more like the female brain than either of them look like the brain of another species," Ruben Gur admits. But this similarity aside, he's nevertheless convinced that women's brains are different in a host of other ways, and that this in turn reveals something about how women think and behave. "The whole brain volume is in keeping with the body size, but the composition of tissue within the brain is different, with females having a higher percent of gray matter and men having higher percent of white matter," he tells me.

Upon this observation lies the latest battleground in the gender wars. Having failed to show that brain size makes any difference, scientists like the Gurs have instead turned their attention to composition.

―――――

A cross-section of the human brain looks something like a freshly cut cauliflower. At the flowery ends are pinkish gray areas, known as the "gray matter." This is what we generally think of as the energy-consuming workhorse part. In the gray matter, the bodies of brain cells translate chemical signals into electrical messages that can travel through the brain, helping it to take care of functions such as muscle control, seeing, hearing, remembering, speaking, and thinking. This is why people sometimes use the terms "brain" and "gray matter" interchangeably.

But there's more to the brain than the tasty, flowery ends of the cauliflower. At the woody stems is the white matter, containing the thin, stringy tails of the brain cells, which make longer-distance connections to and from different parts of the brain. Using the connections in white matter to understand the brain's architecture is a fairly new trend in neuroscience. It can be thought of as figuring out how a radio works not just by looking at the transistors but also by studying the wiring.

This work has been helped by a fairly new technique for scanning the brain, called "diffusion tensor imaging," which allows researchers to picture the strength of the connections in these wires. Neuroscientist Paul Matthews at Imperial College London tells me, "It has completely changed the game, because it allows observations to be made at scale. You can look at the whole brain very rapidly." Observations that would once have taken years can now be done in an afternoon. And it's this technology that Ruben Gur and Raquel Gur, along with a large team of colleagues, used in an important study, published in the January 2014 issue of *Proceedings of the National Academy of Sciences*, looking at how women's brains are wired differently from men's.

Their paper stood out among the hundreds, even thousands, of studies into sex differences that get published every year. One reason for this was that the team studied a large group of people, almost a thousand, between the ages of eight and twenty-two. This size helped to lend it greater scientific value. Second, the findings were dramatic. A study he did in 1999, explains Ruben Gur, showed that in males, "a much higher percentage of the brain is devoted to white matter." Meanwhile, "females have the same volume, or even greater volume, of corpus callosum, which is the largest body of white matter, the nerve fibers that connect the two hemispheres." This new 2014 study went beyond volume to investigate the strength of the connections inside these two areas of white matter. And it seemed to confirm that men have more connections *within* the left and right halves of their brain, while women have more connections *between* the two of halves of their brain.

One popular feature of their research was the pictures. Their published paper was peppered with dazzling images of brains overlaid with blue, orange, green, and red lines to indicate how strong some of these pathways are. One image in particular, which has been reprinted by newspapers and websites all over the world, shows a male brain crisscrossed by blue lines within the hemispheres, and, beneath it, a female brain with orange zigzags showing a dense cluster of wiring between the two hemispheres. It made for perfect headline material, appearing to be nothing less than a literal representation of how differently the sexes think.

When the paper came out, the *Atlantic* magazine immediately declared, "Male and Female Brains Really Are Built Differently," while the *Telegraph* newspaper, in the United Kingdom, announced, "Brains of men and women are poles apart." Not entirely convinced, the online magazine the *Register* went with a tongue-in-cheek headline: "Women crap at parking: Official."

What really captured the world's attention was what the scientists suggested their data might tell us about how men and women behave. An earlier behavioral study the Gurs and their colleagues carried out on the same group of people, published in 2012, claimed to see "pronounced sex differences, with the females outperforming males on attention, word and face memory, and social cognition tests, and males performing better on spatial processing and motor and sensorimotor speed." They argued that their new wiring diagrams, produced using the power of diffusion tensor imaging, could explain some of these differences.

"You need white matter in order to do spatial processing. It requires a lot of interconnectivity between regions to create a three-dimensional object and be able to rotate it in your mind in different directions," explains Ruben Gur. This, apparently, is a feature of the male mind. "Males will have an easier time seeing and doing." When I press him on what this means in practice, he tells me that they might react faster to what they see. For instance, if a man spots a lion about to attack, he might run away more promptly. In females, meanwhile, he sees links between the "verbal, analytic" parts of the brain and the "spatial, intuitive" parts. "I think, for women, they may have an easier time putting together their verbal thoughts with their intuition. If they are more intuitive, then they will be able to articulate the intuition better, at least to themselves," he speculates a little vaguely.

At the time the paper was released, the media were aided by a press release sent out by the University of Pennsylvania's medical school, designed to translate the findings into terms the public might better understand. This release made claims that went far beyond what the paper actually said. It stated that the brain-wiring differences shown by the Gurs and their colleagues indicate that men are better at carrying out a single task while women are better at multitasking. Ruben Gur himself admits to me that he hasn't seen any scientific evidence to support this claim, and he's not sure how it made it into the press release.

But at the time, when researchers spoke to reporters, they went even further. One of the paper's coauthors, Ragini Verma, an associate professor working in biomedical image analysis at the University of Pennsylvania, told the *Guardian*, "I was surprised that it matched a lot of the stereotypes that we think we have in our heads." She added, "Women are better at intuitive thinking. Women are better at remembering things. When you talk, women are more emotionally involved—they will listen more." She told the *Independent*, "Intuition is thinking without thinking. It's what people call gut feelings. Women tend to be better than men at these kinds of skill which are linked with being good mothers."

Characterizing the sexes in this way is sometimes euphemistically phrased as women and men "complementing" each other. Different but equal. They're useful in their own ways, just not at the same things. It's an idea that runs through some religious texts, but was also popular during the Enlightenment in Europe, as thinkers then grappled with how a woman's role in society should be defined. The eighteenth-century French philosopher Jean-Jacques Rousseau was among many intellectuals—male and female—who argued against women's equality on the basis that they weren't the same physically or mentally, but each designed for their own separate spheres. The notion of complementarity thrived through to the Victorian era and ultimately became epitomized in the 1950s middle-class suburban housewife. She fulfilled her natural role as wife and mother, while her husband fulfilled his role as breadwinner.

According to Ruben Gur, his findings reinforce this idea that women complement men. "I'm impressed by the complementarity between the sexes," he replies, when I question him on what his results tell us about the brain. "It almost looks like what is strong in one sex will be weaker in the other and whatever that difference is in the other sex you'll find a complementary effect in the other. Biologically, we are built to complement each other."

"I think they have a particular mission."

"This is an eighteenth-, nineteenth-century problem. We really shouldn't be talking in these terms. I don't know why we're still doing it," complains Gina Rippon, professor of cognitive neuroimaging at Aston University in Birmingham in the United Kingdom. Her long, narrow office, in what proudly claims to be one of the biggest freestanding brick buildings in Europe, is scattered with books on neuroscience and gender. On the shelf are a couple of tiny replica brains and a white coffee cup shaped like a skull. She is one of a small but growing number of neuroscientists, psychologists, and gender experts scattered across the world who are desperately batting away claims that brains show significant sex differences. In the twenty-first century, she is fighting Helen Hamilton Gardener's old war.

Rippon became interested in sex and gender when she was teaching courses on women and mental health at the University of Warwick, where she spent twenty-five years at the start of her career. More women than men tend to suffer from depression or have eating disorders, and she found that, time and again, their illnesses were being explained as something innate to them as females that made them vulnerable. She was instead convinced that there were stronger social reasons for their mental problems. This sparked a fascination with how biological explanations are used and misused, particularly when it comes to women.

"That's the point I was called a feminist biologist," Rippon tells me.

When she arrived at Aston University in 2000 and started working in neuroimaging, she decided to take a look at how the latest powerful imaging techniques were being used in research on women. Technologies like electroencephalography had already been used for almost a century to study electrical signals in the brain. But during the 1990s, functional magnetic resonance imaging—a technique that allows researchers to track changes in brain activity by measuring which areas see more blood flow—utterly transformed the field. There was an explosion of new studies, many of which came tagged with eye-catchingly colorful pictures of the brain.

With this, Paul Matthews of Imperial College London informs me, "cognitive neuroscience was born." It became the most popular way of watching what happens to brain activity when people carry out different tasks or experience emotions.

Despite the promise of this new technology, however, the pictures it painted weren't always pretty. Especially for women. "I did a review in 2008 of where we were going with the emerging brain imaging story and gender differences, and I was horrified," says Rippon. Studies, including some carried out by Ruben Gur at the University of Pennsylvania, saw sex differences in the brain when it came to almost everything. Examples included verbal and spatial tasks, listening to someone read, responding to psychological stress, experiencing emotion, eating chocolate, looking at erotic photos, and even smelling. One claimed that the brains of homosexual men had more in common with the brains of straight women than straight men.

"I just got drawn into it because I thought this is horrendous, that it is being used in exactly the same way as people in the past saying women shouldn't go to university because it will mess up their reproductive systems," she tells me.

Rippon wasn't the only one raising her eyebrows at some of these brain studies. Functional magnetic resonance imaging produces pictures that can be easily skewed by noise and false positives. The best resolution it can reach is a cubic millimeter or so, and with many machines, it's considerably less. This may sound like a tiny volume, but is in fact vast when it comes to an organ as dense as the brain. Just one cubic millimeter can contain around a hundred thousand nerve cells and a billion connections. Given these limitations, people inside the scientific community began to be concerned that they might be reading too much into brain scans.

All over the world, what started as quiet criticism became a crescendo. In 2005 Craig Bennett, then a first year graduate student at Dartmouth College in New Hampshire, carried out an equipment test that inadvertently revealed how it might be possible to read just about anything into a brain scan. He and a colleague tried to find the most unusual objects they could fit inside a functional magnetic resonance imaging machine, to help calibrate it before their serious scientific work began. It was a joke that started with a pumpkin and ended with a dead, eighteen-inch-long, mature Atlantic salmon wrapped in plastic. A few years later, when Bennett was looking for evidence of false positives in brain imaging, he dug out this old scan of the salmon. By chance, proving the critics right and showing how even the best technologies can mislead, the picture happened to show three small red areas of activity close together in the middle of the fish's brain. The *dead* fish's brain.

Amusing though the salmon experiment was, it highlighted what some saw as a far more serious problem in neuroscience. Eight years after Bennett's fish trick, the journal *Nature Reviews Neuroscience* published an analysis of neuroscience studies and reached the damning verdict that questionable research practices were leading to unreliable results. "It has been claimed and demonstrated that many (and possibly most) of the conclusions drawn from biomedical research are probably false," the article began.

The authors explained that one big complication is that scientists are under enormous pressure to publish their work, and journals tend to publish results that are statistically significant. If there's no big effect, a journal is less likely to be interested. "As a consequence, researchers have strong incentives to engage in research practices that make their findings publishable quickly, even if those practices reduce the likelihood that the findings reflect a true . . . effect," they continued. They pointed out that "low statistical power" was an "endemic problem" in neuroscience. In summary, scientists were being pressured to do bad research, including using small samples of people or magnifying real effects, so they could appear to have sexy results.

Paul Matthews, a highly respected British neuroscientist, admits that in the early days of functional magnetic resonance imaging, many researchers—himself included—were caught out by unintentionally bad interpretations of data. "The errors that have been made have been fundamental statistical errors. We've all made them," he says. "I'm more careful about it now, but I've made them, too. It's a very embarrassing thing. It's born of this strong drive to derive results from whatever works one's completed, because one can't do anymore. . . . Most people, if not the overwhelming majority, don't intend to cheat. What they try to do is get excited because of exploration, and they misstate the degree to which they're exploring the data or the meaningfulness of the exploratory outcomes."

The problem has at least been recognized. Even so, Gina Rippon believes that sex difference research continues to suffer from bad research because it remains such a hot-button topic. For scientists and journals, a sexy study on sex difference can equal instant global publicity.

The vast majority of experiments and studies show no sex difference, she adds. But they're not the ones that get published. "I describe this as an iceberg. You get the bit above the water, which is the smallest but most visible part, because it's easy to get studies published in this area. But then there's this huge amount under the water where people haven't found any differences," Rippon explains. People end up seeing only the tip of the iceberg—the studies that reinforce sex differences.

Ruben Gur and Raquel Gur have contributed a sizeable chunk of work to the visible tip of the iceberg, she says. "I think they have a particular mission."

In her 2010 book *Delusions of Gender*, psychologist Cordelia Fine coins the term "neurosexism" to describe scientific studies that fall back on gender stereotypes, even when these underlying stereotypes are themselves un-proven. Ruben Gur's 2014 study on sex differences in white matter between men and women, Gina Rippon tells me, is among those that deserve to be described as "extremely neurosexist."

"Ruben Gur's lifelong passion is to investigate, enumerate, identify, and prove that there are sex differences in the brain," she continues. "A very strong belief in psychological sex differences and explaining them in terms of brain characteristics. That's his life's work, and his lab is still producing that material. It's an impressive body of work, but it's not until you start drilling down into it, in quite an arcane fashion in some cases, that you see that actually some of it is quite flawed."

Critics have questioned, for instance, the Gurs' underlying assumption that men and women perform differently when it comes to social cognition tests, spatial processing, and motor speed. Study after study has shown almost all behavioral and psychological differences between the sexes to be small or nonexistent. Cambridge University psychologist Melissa Hines and others have repeatedly demonstrated that boys and girls have little, if any, noticeable gaps between them when it comes to fine motor skills, spatial visualization, mathematics ability, and verbal fluency.

When it comes to the paper on white matter, Rippon explains, every sex difference that Ruben Gur and his colleagues claim to see can be accounted for by the fact that men have a larger body size and brain volume. As the brain gets bigger, other areas have to get bigger too, in different proportions depending on what's important to keep the brain functioning normally. "If you look at it as a scaling problem, the gray and white matter will change as a function of the brain size, so even that is to do with size."

Others have pointed out that the Gurs were never clear about the true magnitude of the statistical effect and how significant it actually is. "What proportion of all connections are different is a question they didn't really address," says Paul Matthews. Some have even accused the Gurs and their colleagues of cherry-picking the handful of possible pathways, among many, that happen to show some sex difference and using those selectively in their blue and orange illustrations of the brain. This also assumes that all the chosen pathways are being actively used, says Rippon, which isn't necessarily the case.

"It assumes there is this dichotomy between males and females, that we're completely separate," she adds. This is sometimes described as our brains being "sexually dimorphic," meaning that they take two completely different forms in the same species—in the same way as a penis and vagina are dimorphic body parts. Judging by the final, dazzling pictures, the differences looked huge. Neuroscientist and Tel Aviv University professor Daphna Joel echoed this complaint in a letter to the *Proceedings of the National Academy of Sciences*, which published the Gurs' original paper on white matter. "No wonder the main message the reader is left with is of a 'male brain' and a 'female brain' that seem to have been taken from subjects from different galaxies, not just from different planets," she wrote.

Certainly, more recent studies suggest that sex differences in parts of the brain are not as big as scientists once thought. A 2016 paper in the journal *NeuroImage* established that the hippocampus—a brain region that many researchers have claimed is bigger in females—is in fact the same size in both sexes. Led by Lise Eliot, an associate professor of neuroscience at Rosalind Franklin University of Medicine and Science in Chicago, researchers analyzed findings from seventy-six published papers, which together studied six thousand healthy people. Their findings helped dispel the assumption, on physical grounds at least, that women must have a stronger verbal memory, have better social skills, and are emotionally more expressive.

Eliot added that this analysis has shown that there's also no difference in the size of the corpus callosum—the very region of white matter that Ruben Gur claims is on average bigger in women.

"Sex differences in the brain are irresistible to those looking to explain stereotypic differences between men and women," she told reporters when her paper came out. "They often make a big splash, in spite of being based on small samples. But as we explore multiple data sets and are able to coalesce very large samples of males and females, we find these differences often disappear or are trivial."

"Science doesn't operate in a political vacuum."

"The criticisms are nonsense, the criticisms are nonsense," argues Larry Cahill, a professor of neurobiology and behavior at the University of California, Irvine. He tells me that attacks on Ruben Gur's work by Gina Rippon, Daphna Joel, and others are "spurious" and "bogus." Sex differences in the brain "range from small to medium to the enormous," he continues. And on the enormous end of the spectrum are the differences in white matter. He doesn't accept that scaling up for brain size alone can account for the variations.

For the last fifteen years, Cahill has been on what he describes to me as a "crusade" to prove that the brains of women aren't the same as the brains of men. "The way I like to put it, it's not an issue I was looking for. It was an issue that found me," he explains. "I was a neuroscientist like any other, happily operating under the assumption that it doesn't make a damn bit of difference whether or not I'm talking about a male or female, outside the very limited brain regions associated with reproduction." Then, in 1999, he discovered a sex difference in the amygdala, an almond-shaped corner of the brain associated with emotional memory. "I published that in the year 2000, and that was a crossing-the-Rubicon moment," he tells me.

When he started the crusade, he was warned by senior colleagues not to wade into what was then seen as politically sensitive territory. But he pressed ahead, nonetheless. "I came out of the womb stubborn, and when I'm convinced that I'm right about something, I tend to say 'Damn the torpedoes! Full speed ahead!' And that's what I did. I'm glad that I did." Studying the literature, he claimed to find "several hundred" papers supporting the idea that there were unexplained sex differences in the human brain. "It's not the case that sex differences only matter for some tiny structures deep in the brain directly related to reproduction. No. Sex differences are *everywhere.*"

He adds that scientists like Ruben Gur are fully entitled to speculate about what their data might tell us about human behavior. "They engage in perfectly reasonable speculation about what these differences might mean.

Just as you and I might engage in perfectly reasonable speculation about what the anatomical differences may mean."

For Gina Rippon, this has become a tiresome battle. "There are people like Larry Cahill who call us 'sex difference deniers,' but it's the same kind of attack that gets put on feminism at each stage, or whatever wave you think you're in," she tells me. "I'm not paranoid or a conspiracy theorist, but there is a very strong, quite powerful backlash in this area. It's kind of acceptable in an odd way, which is not true if you're talking about race or religion." As someone outspoken about sexism in science, she occasionally receives misogynistic e-mails from men who disagree with her. The worst ones attach photos of their genitalia.

Another recent clash was with the British chess grandmaster Nigel Short. In 2015 he wrote a provocative article in a chess magazine trying to explain why there are so few female players. "Men and women's brains are hardwired very differently, so why should they function in the same way?" he asked. "I don't have the slightest problem in acknowledging that my wife possesses a much higher degree of emotional intelligence than I do. Likewise, she doesn't feel embarrassed in asking me to manoeuvre the car out of our narrow garage. One is not better than the other, we just have different skills." When his comments went viral, Rippon was invited to talk about them on the national BBC radio show *Woman's Hour*. "He thinks that there aren't very many women chess players because they *can't* play chess. It's actually that they *don't* play chess," she argued. Female chess players have said that the aggressive, macho, and sexist atmosphere of professional chess can drive them away.

Rippon tells me that in her field it's impossible not to see the scientific data politicized, especially when it enters the public realm. "Science doesn't operate in a political vacuum," she explains. "I think there are some sciences which can be more objective than others. But we are dealing with people, we're not the Large Hadron Collider." Unlike particle physics, neuroscience is about humans, and it has profound repercussions for how people see themselves.

"It's not something that people don't know much about. This is about everybody's lives. Everybody has a brain, everybody has a gender of some kind, . . . they've either been in a mixed-sex school or they have worked in a mixed-sex environment. They've got boys and girls. So they see differences. And so, when you say there aren't really any, they say you're wrong," she adds. She has seen it for herself when giving lectures about her work. "I go into schools and talk to girls, and their whole expectation is far more gendered than it used to be. These are toxic stereotypes and these girls' futures are being affected by this."

According to social psychologist Cliodhna O'Connor based at Maynooth University in Ireland, Ruben and Raquel Gur's study on white matter is a textbook example of how research into sex differences can quickly become absorbed into people's wider gender stereotypes. When the paper was published in 2014, she decided to monitor reaction to it. What she found was shocking. "It was covered in all the major national newspapers," she tells me. "The main meaning that was taken out of it was just the fact that men and women are fundamentally different in some very essential, primitive, unavoidable way."

O'Connor found that people in the thousands commented online and discussed the research on social media such as Twitter and Facebook. "As a conversation evolved, cultural and gender stereotypes were progressively projected onto that scientific information, to the extent that people were describing the research as the discovery of stuff that wasn't even mentioned in the original scientific article," she says. People latched on to the idea in the press release, but not in the paper, that women are better at multitasking. Before long, they were using the study to argue that men are more logical while women are more emotional. "That dichotomy wasn't mentioned either in the press release *or* the original article, but it was kind of spontaneously introduced when people were discussing the research," she adds.

O'Connor tells me that this kind of distorted reaction to brain studies on sex and gender is common. "No matter how neutral the initial presentation of information, people do tend to gradually recruit the stereotypes and the associations that are prevalent in a culture and then project that," she explains. It's part of being human. We tend to interpret new information by categorizing it, using whatever understanding we already have, even if this is prejudiced.

Another factor that prompts people to behave this way is that we like to justify the social system we're in. If everyone around us thinks that women are less rational or worse at parking, even the thinnest piece of information that reinforces that assumption will be pasted into our minds. Research that confirms what appears to be obvious seems right. Anything that contradicts it, meanwhile, is dismissed as aberrant. This is why, when theories come along that challenge gender stereotypes, we may also find them more difficult to accept.

But all this still leaves one unanswered question: If the brains of women and men aren't so different, then why do researchers like Ruben Gur and Larry Cahill keep seeing sex differences?

"If you take any two brains, they are different."

At the turn of this century, Londoners were surprised by a revelation about one of their most recognizable groups of workers. The brains of the city's black-cab drivers, who are famous for their perfect navigational ability, down to the smallest and most hidden side streets, were being physically altered by their work.

Neuroscientist Eleanor Maguire at University College London discovered that the mental feat of memorizing the layout of twenty-five thousand streets and thousands of landmarks, known as "The Knowledge," could be changing the size of a cabbie's hippocampus, a region associated with memory. This piece of research had enormous implications. It helped confirm an idea that scientists had already been developing since the 1970s, particularly through animal studies: that the brain isn't set in stone in childhood but is in fact moldable throughout life.

"These changes are terribly tiny, but they are measurable," says Paul Matthews. Studying musicians, basketball players, ballet dancers, jugglers, and mathematicians has confirmed that brain plasticity is real. In the context of sex difference research, it also raises an important question: If intense experience and learning a new task can shape a person's brain, could the experience of being a woman shape it as well? Could plasticity therefore explain the sex differences that are sometimes seen in the brain?

According to Gina Rippon, psychologist Cordelia Fine, and gender scholars Rebecca Jordan-Young in New York and Anelis Kaiser in Bern, Switzerland, plasticity is a phenomena that has been oddly ignored when people talk about sex differences in neuroscience. "Our brain actually absorbs a lot of information all the time, and that includes people's attitudes to you, expectations of you," says Rippon. Her own work is driving her toward the view that it isn't just supreme feats of learning or traumatic experiences that affect the brain but more subtle and prolonged things, too, like the way girls and women are treated by society.

This idea has in turn been woven into an even bigger and more radical new theory that might explain how the small sex differences we occasionally see in brain composition might emerge. Rippon, Fine, Jordan-Young, and Kaiser have argued that biology and society are "entangled"—that they work in concert with each other, through mechanisms like plasticity, to create the complicated picture we call "gender."

Their ideas are supported by a growing body of evidence on how gender differences shift over time. Research in the 1970s and 1980s revealed that the number of American boys with exceptional mathematical talent outnumbered girls by thirteen to one. At the time, it was seen as a shocking imbalance. Since then, however, as psychologists David Miller and Diane Halpern (Halpern is a former president of the American Psychological Association) have pointed out, this ratio has plummeted to as low as four, or even two, to one. In a paper published in 2014 in *Trends in Cognitive Sciences*, they note that there have been equivalent drops in gaps in general performance on math tests in the United States.

But how? If mathematics ability were rooted in biology and sex differences were fixed, then we wouldn't expect to see these changes over time. What's more, we would expect the differences to be the same everywhere. And they're not. Among Latino children in American kindergartens, for example, girls tend to be the best achievers in mathematics tests, not boys. "Challenging the notion of universal male advantage in mathematics, sex differences in average mathematics test performance are not found in many nations and are even reversed (female advantage) in a few," Miller and Halpern observe. What looks like a biological difference in one particular place and time can turn out to be a cultural difference after all.

Plasticity and entanglement suggest that, like London cabbies memorizing street layouts, culture can have a ripple effect on biology. We know, for instance, that playing with certain toys can actively affect a child's biological development. "We're good at what the brain allows us to be good at and, as we become good at something, our brain changes to enable that," explains Paul Matthews. Playing action video games or with construction sets, for instance, improves spatial skills. So if a young boy happens to be given a building set rather than a doll to play with, the stereotype of males having better spatial skills is physically borne out. Society actually ends up producing a biological change.

On the flip side, exposing someone to bad stereotypes can impair their performance. In one controversial study that Miller and Halpern cite, women who are reminded of negative stereotypes about female abilities in math go on to perform worse on math tests. "Removing stereotype threat can improve both men's and women's academic achievement," they write.

With all these effects on the brain, in a world as gendered as ours, says Rippon, it's actually surprising that we don't see more sex differences in the brain than we do. But then, so many factors other than our gender affect us. Plasticity and entanglement imply that every single brain must be unique, for the simple fact that every person's life experience is different. It is this, argues Daphna Joel at Tel Aviv University, that makes looking for differences between groups so fraught with error. Evidence of sex difference in the brain is statistically problematic because each brain varies from the next.

This may go some way to explaining why neuroscience and psychological studies often get different results when they're looking at the same thing. If one piece of research doesn't confirm a sex difference where another claims to have found one, scientists sometimes assume that they must have made some mistake and pulled out a false negative. "They have many explanations to explain why they fail to find differences," says Joel. "They never say that maybe there are no differences, and the fact that someone else found a difference is just a chance finding, and it's actually a false positive. It's especially amazing, because in science this is the first thing you need to think, that if you don't find difference, maybe the theory is wrong."

This way of thinking suggests that it's not varying environments, false negatives, or bad experiments that are obscuring evidence of the brains of women and men being sexually dimorphic. It's that there isn't dimorphism in the brain to begin with. "Every brain is different from every other brain," Gina Rippon explains. "We should take more of a fingerprint type of approach. So there is some kind of individual characteristic of the brain, which is true of the life experiences of that person. That's going to be much more interesting than to try to put them all together, trying to squeeze into some kind of category."

Daphna Joel's theory, published online by the *Proceedings of the National Academy of Sciences* at the end of 2015, states that, rather than being distinctly male or female, the brain is a unique "mosaic" of characteristics. In any given person, you're likely to find features in a form that's more prevalent in men and also in a form that's more prevalent in women. To illustrate, she gives me the example of porn and soap operas. Watching porn is an interest strongly associated with men, but not all men watch porn, and of those who do, some also love watching soap operas, which is an interest commonly associated with women. Add up all the different interests that overlap and, in any person, you're left with a huge gender mix. "Of course, most of the features will just be in an intermediate form, something that is common in both males and females," she adds.

The idea of exploring a range of features across the brain as a whole came as a revelation to Joel. It was sparked by studies reporting that environmental factors can reverse the effects of some sex differences in rats. "Regardless of how stressed your mother was when she was caring for you, where you live, or what you eat, your genitals will not change. The sex effects on the genitals are fixed, always the same. But when I saw that the sex effects on the brain can actually be opposite, so what you see in one sex under some conditions, you can see in the other sex under other conditions, I realized that I was using sex effects on the genitals as an implicit model when thinking about sex effects on the brain," she tells me. "This is not a good model."

Researchers rarely look at the brain in this way. Very often they're studying just one brain region, like the amygdala or hippocampus, or one particular behavior, like mathematical ability or watching porn. Looking at the brain and behavior as a whole produces very different results when it comes to sex difference. Joel's research reveals that, depending on the study, between 23 and 53 percent of people show variability in their brains, with features associated with both men and women. Meanwhile, the proportion of people in the studies she has analyzed that have purely masculine or purely feminine brain features is between none and 8 percent.

"If you take any two brains, they are different, but how they differ between any two individuals, you cannot predict," she explains. By this logic, there can't be any such thing as an average male or average female brain. We are all, each one of us, a mix. Our brains are intersex.

––––––––

Having the fresh perspective of female researchers like Gina Rippon, Anne Fausto-Sterling, Melissa Hines, Cordelia Fine, and Daphna Joel—while it may not immediately change how science tackles sex difference—may at least force a rethink of old beliefs that women's minds must be essentially different from men's because the only alternative would be that they're identical. They take the black and white divisions of the past and reveal that the truth is more likely to be gray.

Anne Jaap Jacobson, a philosopher and emeritus professor at the University of California, Berkeley, now based at the University of Houston, has coined the word *neurofeminism* to describe this alternative approach to brain science, which attempts to root out stereotypes and look at brains objectively. "A lot of the research starts off with the assumption that various people call 'essentialism.' That men and women are essentially different, that the differences are really sort of basic," she tells me.

"The problem with this question of difference and similarity is that we're all different and we're all similar," explains Daphna Joel. "When people want to study sex in the brain, they immediately translate this to studying sex differences. But already here they make many assumptions, and the first is that there are two populations of brains, male and female. This is an assumption that needs to be shown scientifically, or proven. They say: 'This is solid ground, and from here I continue.' I question the solid ground."

Neuroscientist Paul Matthews agrees that this approach could be a useful corrective for neuroscience. "Comparing males and females at any one time point is a complicated question to make meaningful, because it is actually so ill-defined as posed," he says. "There's a lot of variability in individual brains. In fact, the anatomical variability is much greater than we ever realized before. So the notion that all people of the male sex have a brain that has fixed characteristics that are invariant seems less likely to me. In fact, so much less likely that I think the notion of trying to characterize parts of the brain as more male-like or more female-like actually isn't useful."

Ruben Gur, while he refuses to shift in his conviction that sex differences in the brain are the rule rather than the exception, admits to me that these days he has changed the language he uses. "A lot of people are using the term 'sexual dimorphism' when they talk about sex differences in brain structure, and I'm guilty of that myself," he says. "I've done that, but I don't do that anymore. Because if you think of it, when you talk about dimorphism, you're really talking about sexual dimorphism. You're talking about different forms. So a penis versus the vagina, that's a sexual dimorphism. Having breasts is a dimorphism. I wouldn't go so far as saying the brain is dimorphic. I would say there are some significant differences, sex differences, in brain anatomy, but I wouldn't say they rise to the level of being dimorphic."

———————

All this work on sex difference in our bodies and brains has a story underneath it.

When neuroscientists like Ruben Gur and Simon Baron-Cohen claim to see profound gaps between women and men, they are aware that these gaps aren't spontaneous. If they're there, they're there for a reason. Gur has described them as revealing how we are "built to complement each other," suggesting that humans must have evolved with some sexual division of labor between them. Women, he implies, are the more empathic, intuitive sex, perhaps built for parenting. Men are better at seeing and doing, he says, which seems to suggest they are the natural hunters and builders. Baron-Cohen also argues that men tend to be the systemizers while women are the empathizers.

"If your job is to lift a hundred and fifty pounds, and you can't do it, why should you want to work in that job?" Gur asks me.

It's hard to argue with that kind of logic. But while he's willing to speculate on what biology tells us women have evolved to be able to do, it's a question that's beyond his job description. It belongs to the field of evolutionary biology.

The evolutionary perspective reminds us that our bodies weren't created yesterday. They were forged over millennia, every part slowly adapting to the pressures of the environment to better serve some need. From breasts and vaginas to brain structure and cognitive ability, for every difference or similarity we see, there must be some evolutionary purpose to it. This is where the sex differences and similarities that biologists claim to see in our bodies and brains connect with the story of our past. If women are better survivors than men, the explanation for it is in this tale. If women and men have quite similar brains, the reasons for that are here, too.

Evolutionary biologists have the almost impossible task of deciphering this story. Did the sexes complement each other, the way Ruben Gur suggests, or did they do the same jobs and share parenting? Were women crouched around a campfire, tending to children and waiting for male hunters to bring home the bacon? Were they independent, hunting for their own food? Were they monogamous or promiscuous? Were males always dominant over females?

They are questions that science may never fully answer, but there are ways to try. One window on the past is provided by primatologists studying our closest animal cousins, the great apes, from whom humans split around five million years ago. Studying how they interact gives us some insight into what our basic way of life may have been before we became the species we are now. Another window comes from evolutionary psychologists, who try to picture life in the Pleistocene, the epoch during which modern humans evolved looking anatomically the way we do now. Then there's archaeological evidence, such as tools and bones. By observing the lives of modern-day hunter-gatherers, anthropologists can also draw portraits of how early woman might have lived.

Writing our evolutionary story isn't easy and it's also plagued by controversy. As Charles Darwin's work in the nineteenth century proves, the narratives have often been shaped by the attitudes of the time. Even he, the father of evolutionary biology, was so affected by a culture of sexism that he believed women to be the intellectually inferior sex. It's taken a century for researchers to overturn these old ideas and attempt to rewrite this flawed tale.

CHAPTER 5

Women's Work

We still live in a world in which a significant fraction of people, including women, believe that a woman belongs and wants to belong exclusively in the home; that a woman should not aspire to achieve more than her male counterparts.

—Rosalyn Sussman Yalow, in her banquet
speech on being awarded the Nobel Prize in
Physiology or Medicine, December 1977

The long road to the sprawling home of Sarah Blaffer Hrdy, primatologist, anthropologist, and emeritus professor at the University of California, Davis, is flanked by dry fields. She and her husband carved their walnut farm here, near Sacramento, out of almost nothing. The trees are new, the pastures on which their lambs and goats are feeding are new, and they planted the spindly silver walnut groves themselves. She survives under the looming possibility that wildfires could come along to claim it all, as they've almost done in the past.

But then any fire would have to battle Hrdy herself, who, now seventy years of age, is a force of nature in her own right. Hrdy's work into what primate behavior can tell us about human evolution, one scientist tells me, reduced her to tears. For Hrdy's groundbreaking ideas on women, she's been described as the original Darwinian feminist.

Primatology is today a female-dominated field, guided by early pioneers like Jane Goodall and Dian Fossey. But when Hrdy started her career in the 1970s, not only did men rule the roost, the accepted wisdom was that human evolution had been shaped largely by male behavior. Males were the ones under pressure to attract as many mates as possible to increase their odds of having more offspring, males were aggressive and competitive in their quest for dominance, and males needed to be creative and intelligent when they hunted for meat.

As our closest evolutionary cousins, primates were naturally expected to follow similar patterns. When male primatologists went into the field, they would often focus on aggression, dominance, and hunting, Hrdy tells me. Females were routinely overlooked. They were believed to be passive, sexually coy, and generally at the mercy of stronger, larger males. Indeed, early studies of chimpanzees—a species in which males happen to be particularly aggressive and dominant—reinforced this.

Things changed for Hrdy when she went out into the field for herself. She finally saw how this account of females might be wrong.

It began with a trip to Mount Abu, a region of Rajasthan, northwestern India, which is home to a species of monkey known as the *Hanuman langur*. Hanuman is the name of the Hindu monkey god, a symbol of strength and loyalty, while "the name *langur* is Sanskrit for having a very long tail," she explains to me in her large office, which is decorated with framed drawings of primates. "They are the beautiful, elegant gray ones with the black gloves and faces." Hrdy had heard that male langurs were killing infants of their own species. It was so strange a phenomenon that scientists assumed there must be something desperately wrong with them. Animals simply didn't behave in ways that were bad for their group, they thought. The only possible cause must be that the male monkeys had gone mad. Overcrowding had created a pathological hotbed of aggression, perhaps.

The truth was stranger. When Hrdy watched closely, she began to realize the murders weren't random acts of madness at all. In the everyday course of life, she noticed that male langurs were far from violent toward infants. "I would see young langurs jumping on a male langur reclining on the ground as if he were a trampoline. He was completely tolerant of the infants in his troop. There was nothing pathological about it," she explains.

The rare infanticides turned out instead to be carefully calculated. And they were committed by males from outside the breeding group. "When I first did see infants missing, and then later I actually saw a male attack infants, it was very goal-directed stalking, as if by a shark. Day after day, hour after hour." What was making a male commit this gruesome killing was the expectation that, without her baby, a mother would have to mate again. If he didn't kill the infant, he would have to wait a year before she finished nursing and started ovulating. She couldn't mate any sooner.

To scientists, the idea was shocking. Hrdy had shown that a monkey could choose to kill a healthy young member of his own species simply to perpetuate his bloodline. Infanticide went on to become a fruitful area for animal research. The behavioral patterns Hrdy saw, detailed in her 1977 book, *The Langurs of Abu: Female and Male Strategies of Reproduction*, have since been reported for more than fifty primate species as well as other animals.

But something else also fascinated her about these murders. It was the extraordinary way the female Hanuman langurs reacted. They weren't passive. They didn't carelessly allow their infants to be killed by aggressive males. Instead, they banded together and put up fights to fend them off. This observation, too, challenged long-standing ideas about natural primate behavior. It showed that females weren't only fiercely protective of their children (which might have been expected), but that they could also be aggressive and cooperative.

Questioning assumptions can have a remarkable ripple effect. Further work by Hrdy showed that female langurs were promiscuous, too, contrary to popular wisdom about females being sexually coy. Male langurs, she noticed, attacked only those infants being carried by an unfamiliar female—never by a female with which they'd mated. By having as many mates as possible, Hrdy suggested that female langurs might be strategically lowering the odds of a male killing her infant.

It became impossible for primatologists to ignore females any longer.

———

Hrdy believes that being a woman in her field is one reason she noticed behavior that hadn't been recognized before. She was driven to investigate what others may have chosen to overlook. "When a langur female would leave her group, or when she would solicit a male when she was pregnant, a male observer may say, 'Well, that's just freak,' and not even follow her to find out where she was going or what she was doing. A woman observer might empathize more with the situation or be more curious."

Her work didn't just mark a sea change in how primates were beginning to be understood but was a personal revelation as well. Hrdy had been raised in a conservative, patriarchal family in south Texas. Noticing how competitive and sexually assertive females could be in the rest of the primate world prompted her to question why women in her own society should be thought of as any different. Primates, particularly great apes such as chimpanzees, bonobos, gorillas, and orangutans, have long been used by science as a way of understanding our own evolutionary origins. We share roughly 99 percent of our genomes with chimpanzees and bonobos. In genetic terms, we are so close that primatologists routinely refer to humans as another great ape. So if other female primates could show so much variation in their behavior, why did evolutionary biologists still characterize women as the naturally gentler, more passive, and submissive sex?

Trying to get her male colleagues to see primates from a female's perspective, though, was a battle. When Hrdy returned from her fieldwork in Mount Abu in the 1970s, despite social change happening around her, including a resurgence in feminism, science was still very much a boys' club. One time at a conference, when she was asked to define what feminism meant to her, Hrdy recalls saying, "A feminist is just someone who advocates equal opportunities for both sexes. In other words, it's being democratic. And we're all feminists, or you should be ashamed not to be." But equal opportunities weren't always encouraged, in her field at least. Her work, as well as that of many other women scientists, was sometimes treated differently from that of her male counterparts. Some people refused to acknowledge her research, let alone incorporate its ideas.

Hrdy used to get together with other female researchers at women-only house parties to discuss the problems they faced. They euphemistically called them their "broad discussions." And there was plenty to discuss. The influential evolutionary biologist Robert Trivers, a colleague of Hrdy's, once told a reporter that Hrdy should concentrate on being a mother instead of on her work She forgives him now, she tells me. (Trivers, meanwhile, tells me that he intended the remark to be a secret, and admits he's sorry it was made public.)

Exasperated, she even used her work on apes and monkeys to make covert remarks about her male colleagues. "I was writing about how male baboons were the basis of social organization. Males compete with males, and then the dominant males form alliances with each other so as to improve their access to females. And then I would make these very oblique parallels to what went on in American universities," she remembers. "I was, of course, referring to male professors who, when called out for sleeping with academic subordinates, would back one another up. All through my career, these things were going on."

Hrdy's feminism and science met in the middle, not just because of the behavior of some men in her field but also because she recognized that scientific theories that ignored female behavior were incomplete. "In science, paying equal attention to selection pressure on both males and females, that's just good science. That's just good evolutionary theory," she tells me. One of the most important frontiers, as she saw it, was understanding mothers and how they defined a woman's role in human evolution. It was a question that would also lead her back again to the dark phenomenon of infanticide.

"Cooperative breeding in humans is becoming more and more important."

I'm in the ape enclosure at San Diego Zoo, one of the biggest zoos in the world.

I'm transfixed by a fluffy two-year-old bonobo. She's cheerfully hanging on to her mother's fur as the ape leaps from branch to floor, letting go of her to playfully roll on the ground for a few seconds before quickly returning. I have a two-year-old as well. And the bonobos' behavior reminds me of my own close relationship with my son. In the little bonobo I see a similar mischievousness and even the hint in her of his cheeky smile. They watch each other the same way that we do. The similarities between us are uncanny.

At close quarters like this, I start to understand why humans are sometimes regarded as another great ape, alongside bonobos, chimpanzees, gorillas, and orangutans. But as much as we have in common, there's one important contrast between me and the bonobo mother. In the entire time I'm looking into the glass enclosure, I never see her lose contact with her infant. At no point does the little one fall out of her mother's protectively tight orbit. My son, on the other hand, is already at the other end of the enormous zoo with his father.

Human motherhood is rarely the single-handed job that it is for chimpanzees and bonobos. Of course, this is something most of us know from our own experience as children or parents. When I'm at home in London, my son typically spends half the week being cared for by other people, including his father, grandmother, and nursery staff. Aunts, uncles, and friends step in too, sometimes. When I'm traveling for work, I go days without seeing him. This isn't unusual. Few babies or toddlers get through their early years without ever leaving their mothers' sides.

Primates are different. According to Sarah Hrdy, there are nearly three hundred primate species, and in about half of them you'll rarely see a female ape or monkey out of contact with her child. The infants, in turn, stick close to their mothers, sometimes for years. "Under natural conditions, an orangutan, chimpanzee, or gorilla baby nurses for four to seven years and at the outset is inseparable from his mother, remaining in intimate front-to-front contact a hundred percent of the day and night. The earliest a wild chimpanzee mother has ever been observed to voluntarily let a baby out her grasp is three and a half months," Hrdy notes in her 2009 book, *Mothers and Others: The Evolutionary Origins of Mutual Understanding*. She includes a picture she once took of a female langur who was so attached to her baby that she faithfully carried around its corpse after it died.

Others have made similar observations. "Mothers carrying dead infants is not uncommon in the primate world," confirms Dawn Starin, a London-based anthropologist who has spent decades studying primates in Africa, Asia, and South America. In her research on red colobus monkeys in Gambia, one female "carried her maggot-riddled infant around with her for days, grooming it, sticking it in the crotches of trees so that she could feed without it slipping to the ground, and never letting any of the others touch it." Encounters like these left her with the impression that an infant is treated like an extension of the mother's body, a real part of her, and not a separate being.

For humans, the universal pattern seems to be that mothers are just as protective of their children but not so constantly attached. This isn't something that's true only of modern parents in big cities but everywhere across the world. It really does take a village to raise a child.

For anthropologists trying to get a grip on our evolutionary history, the best case studies are people who live the way our earliest ancestors might have, hunter-gatherers. Modern-day hunter-gatherers are rare and dwindling, drawing a subsistence living off the land, foraging for wild plants and honey, or hunting animals. They're an imperfect window on our past, partly because each community is different from the next, depending on its environment, and also because other cultures have encroached on them over the years and distorted how they live. But by watching their lifestyles and behavior, we can still get some sense of how humans might have lived many thousands of years ago, before societies began domesticating animals and before agriculture.

Some of the most studied hunter-gatherer groups are in Africa, the continent from where all humans originally migrated. This makes them arguably the most reliable source of data for evolutionary researchers. They include the !Kung, bushmen and bushwomen living in the Kalahari desert in southern Africa, the Hadza who live in the Lake Eyasi region of northern Tanzania, and the Efé in the Ituri Rainforest in the Democratic Republic of the Congo. Anthropologist Sarah Hrdy notes that all three of these societies have people who play parental roles to other people's children—known as "alloparents."

She describes this system as "cooperative breeding." In her book *Mothers and Others* she writes, "!Kung infants were held by others some twenty-five percent of the time—a big difference from other apes, among whom new infants are *never* held by anyone other than their mother." Among the Hadza, newborns are held by alloparents 31 percent of the time in the first days after birth. For children under four years of age, people other than their mothers hold them around 30 percent of the time. In central-African foraging nomadic communities, including the Efé, mothers share their babies with the group immediately after birth, and they continue this way. Efé babies average fourteen different caretakers in the first days of life, she adds, including their fathers.

One more difference between humans and apes is how we give birth. Chimpanzee females are known to move away and seek seclusion before they give birth, to hide from predators or others who might harm their newborns (chimps enjoy meat, and have been known to kill and eat infants of their own species). Humans, on the other hand, do exactly the opposite. Expectant mothers almost always have people to help them when their babies are due. In my case, it was an entire team, including my husband, sister, doctors, and a midwife. Anthropologists Wenda Trevathan at New Mexico State University and Karen Rosenberg at the University of Delaware have noted that childbirth is a lonely activity in few human cultures. Helpers are so important that women may even have evolved to expect them, they've argued. Their theory is that the awkward style of delivery of human births and the emotional need that mothers have to seek support during birth may be adaptations to the fact that our ancestors had people aiding them when they delivered their babies.

All this evidence suggests that cooperative breeding is an old and universal feature of human life, not a recent invention. And there are good reasons why. "One of the primary traits that we have is that we're sort of the rabbits of the great ape world," explains Richard Gutierrez Bribiescas, professor of anthropology at Yale University, who has studied the role of fathers in human evolution. "We have very high fertility compared to other great apes, compared to chimpanzees, gorillas, and orangutans. And we tend to produce these very large offspring that require a lot of long-term care," he tells me.

Most primates, meanwhile, will generally wait until the first infant has matured before having the next. A female bonobo would struggle to feed herself and move lithely through the forest if she had to drag around a litter of baby bonobos clinging to her fur.

Two notable exceptions are titis and tamarins, both species of New World monkey in which fathers are extraordinarily involved in child care. Anthropologist Dawn Starin tells me, "When I studied a group of titi monkeys in Peru, the infant was usually carried by the father and spent most of its time with him. The father is completely involved with the rearing of the young. The mother was really just a dairy bar, a pair of milk-secreting nipples." Like humans, titi monkeys are cooperative breeders. Some captive studies on this species, she says, have even suggested that the infant may be primarily attached to the father rather than the mother.

Tamarin monkeys also rely on the efforts of both parents, simply to cope. "With tamarin monkeys, for reasons we don't understand, they twin, and the twins are very large," explains Bribiescas. "So the only way that can be viable is . . . some kind of paternal care. Otherwise it is very unlikely that the mother would be able to support these two very large twins." This support is so vital that tamarins are known to neglect their children if they don't have the help anymore. Sarah Hrdy has noted, according to data from a colony living at the New England Primate Research Center, that when a tamarin mate dies, the infants' survival odds plummet. "There was a twelve percent chance of maternal abandonment if the mother had older offspring to help her, but a fifty-seven percent chance if no help was available," she writes.

Abandonment and neglect like this are rare. In the thousands of hours that scientists have watched monkeys and apes in the wild, very rarely has anyone seen one injure her infant deliberately. Primate mothers may be incompetent sometimes, especially with their first babies, but they hardly ever choose to let their offspring die. This, too—shocking though it may sound—is a feature in which humans again mark themselves out from their evolutionary cousins.

The maternal instinct in humans is not an automatic switch, which is flicked on the moment a baby is born.

This is Sarah Hrdy's radical proposition. All over the world, mothers are known to admit that it takes time for them to fall in love with their babies, while some never do. In some unfortunate cases, mothers deliberately neglect and even kill their newborns. This may seem utterly unnatural. After all, we assume the maternal instinct is as strong and immediate in humans as it is in any other creature. It's considered a fundamental part of being a woman. So much so that those who don't want children or reject their own are sometimes considered odd. But the reality, observes Hrdy, is that it's more common for mothers not to form an immediate attachment to their offspring than we like to believe.

Her argument is that this is a legacy of cooperative breeding. Like tamarin monkeys, humans often rely on help to cope with raising their children. Hormones released in pregnancy and childbirth help a mother bond to her baby. But this bond may also be affected by her circumstances. If her situation is particularly dire, she may feel she has no choice but to give up altogether.

In Britain, studies estimate that between thirty and forty-five babies are killed every year—about a quarter of these within the child's first day of life. According to research in 2004 by Michael Craig, a lecturer in reproductive and developmental psychiatry at the Institute of Psychiatry at King's College London, this is likely to be an underestimate, because these kinds of killings can easily go unreported. But even as the reported figures stand, infants are at a bigger risk of homicide than any other age group. For the babies killed soon after birth, the most common perpetrators are teenage mothers, especially those who are single and living at home with parents who might be disapproving of their pregnancies. Most of them aren't killing their babies because they're psychotic or mentally ill, says Craig, but because of the desperate positions they find themselves in.

To make her case, Sarah Hrdy has also investigated a particularly grisly historical example. In the eighteenth century in urban parts of France, as many as 95 percent of mothers sent their children away to be wet-nursed by strangers, sometimes in questionable conditions. Her research, outlined in a series of lectures she gave at the University of Utah in 2001, suggests that the mothers must have known this would dramatically lower their babies' odds of survival. Culture dictated that they do it, so they did. The deadly practice was evidence, she argues, that not every human mother protects her newborn at all costs. Female infanticide in Asia today is sometimes also carried out with the complicity of mothers. Again, society influences how they respond to a birth.

Hrdy's hypothesis about the profound importance of cooperative breeding is a difficult one to prove, especially given the myriad pressures that pregnant women experience in the modern world. But it also has the power to release women of the guilt they may feel when they're unable to cope

alone. If we are natural cooperative breeders—a species in which alloparents are part of the fabric of families—it's unreasonable to expect women to
manage without any help. For Hrdy, a feminist, this line of research also has
obvious political implications. It reinforces why lawmakers shouldn't outlaw
abortion and force women to have babies they feel they cannot raise or do
not want. It also highlights how important it is that governments provide
better welfare and child care for mothers, especially those who don't have
support at home.

The weight of evidence does at least seem to be in favor of the idea that
humans didn't evolve to raise their children single-handedly. Child care was
not the sole responsibility of mothers. "What we're finding is that cooperative breeding in humans is becoming more and more important in terms of
our thinking," agrees anthropologist Richard Bribiescas. As evidence builds
around this and what it means, it's becoming clearer just how important alloparents are in the human story. And it also raises an interesting question:
If mothers didn't evolve to parent alone, who else around them would have
been providing the most support?

> *"We see a huge range of plasticity in how much engagement there is
> in human males."*

Sarah Hrdy tells me that when she welcomed her first grandchild last year,
she took the opportunity to run a small experiment on her family. Arriving at
her daughter's house, she took saliva samples from herself and her husband.
She took another set of samples after spending time with the new baby. Tests
revealed that they had both experienced a rise in oxytocin, the hormone associated with love and maternal attachment.

Our bodies betray how strong the emotional connections can be between children and people who aren't their parents. Physical contact with
a baby, scientists have long known, can have dramatic effects on a mother's
hormone levels. These hormones in turn influence how she bonds with her
child. Others who aren't mothers, we now know, can experience these hormonal changes, too.

Evolutionary biologists have often assumed in the past that, of all the
people providing support to mothers, fathers would have been front and
center. In his 2006 book, *Men: Evolutionary and Life History*, Richard Bribiescas suggests exactly this. And from the perspective of how we've lived for
centuries, often in monogamous marriages and nuclear families, this seems

to make sense. Even if they weren't directly involved in child care, the material help that fathers brought to families, such as food, must have been crucial to keeping children alive and thriving.

Some recent studies, however, don't agree. In a 2011 paper in *Population and Development Review*, Rebecca Sear at the London School of Hygiene and Tropical Medicine and David Coall at Edith Cowan University in Australia pulled together all the published studies they could find on how the presence of fathers, grandparents, and siblings affect a child's survival. They found that other family members were so valuable that, once a child passed the age of two, they could even cushion the impact of an absent mother. Where this help came from, though, was more of a surprise. Older siblings had a more positive effect than anyone besides the mother. After this came grandmothers, then fathers, followed far behind them all by grandfathers.

"Fathers were rather less important: in just over a third of all cases did they improve child survival," Sear and Coall note in their paper.

This doesn't mean that hands-on fathering isn't important. Just that it isn't always there. In 2009 anthropologist Martin Muller at the University of New Mexico and his colleagues studied how much effort men in two neighboring but different East African communities put into parenting. In one, the Hadza hunter-gatherers, they found that fathers were involved in everything from cleaning to feeding infants, spending more than a fifth of their time interacting with children under three if they were in the camp at the same time, and also sleeping close to them. In the other, a pastoralist and warrior society called the Datoga, they found a strong cultural belief that looking after children was women's work, with men eating and sleeping separately and not interacting much with infants. Their hormone levels reflected the difference in parenting styles. The more involved fathers—the Hadza—produced less testosterone than Datoga fathers.

"We see a huge range of plasticity in how much engagement there is in human males," admits Richard Bribiescas, from "the most doting and caring father, and everything is great and lovely, to a father that's sort of engaged and maybe just brings food and resources home, to the ultimate, very horrific cases of things like infanticide." If society expects men to be involved in child care, they are, and they can do it well. If society expects them to be hands-off, they can do that, too.

This plasticity is unique to humans. "In other great apes and other primates you simply don't see that. They're locked into one strategy," he adds.

If in our evolutionary history, caring for children is something that would have been done not just by mothers but also by fathers, siblings, grandmothers, and others, the traditional portrait we have of family life starts to crack. A nuclear family with one hands-on father certainly isn't the norm everywhere. In a few societies, for example, children even have more than one "father." In Amazonian South America, there are communities that accept affairs outside marriage and hold a belief that when a woman has sex with more than one man in the run-up to her pregnancy, all their sperm help build the fetus. This is known by academics as "partible paternity." Anthropologists Robert Walker and Mark Flinn at the University of Missouri and Kim Hill at Arizona State University, who have confirmed how common partible paternity is in the region, claim that children benefit from these family arrangements. With more fathers, their odds of survival go up. They have more resources and better protection from violence.

This all points to the possibility that living arrangements among early humans could have taken any number of permutations. Monogamy may not have been the rule. Women, if they weren't tied to their children all the time, would have been free to go out to get food and perhaps even hunt. The Victorian ideal that Charles Darwin based his understanding of women upon—mother at home, taking care of the children, hungrily waiting for father to bring home the bacon—is left out in the cold.

"A theory that leaves out half of the human species is unbalanced."

It was April 1966.

Some of the most important names in anthropology had come together at the University of Chicago to debate what was then a fast-growing body of research about the world's hunter-gatherers. The symposium they were all a part of was headlined "Man the Hunter." And they would help shape the way a generation of scientists thought about human evolution.

The gathering was appropriately titled. The "man" in the title, as anyone attending would have guessed, really did refer to men, not to all humans. In almost no hunter-gatherer communities were women known to routinely hunt. Even so, this one activity was believed to be the most important in human evolutionary history. Hunting made men band together in groups and work cooperatively, so they could target their prey more effectively. It forced men to be inventive and create stone tools. Hunting may also have been what prompted men to develop language so they could communicate more effectively. By bringing home meat, followed the logic, men were able to provide themselves, women, and their hungry children with the densely packed nourishment they needed to develop bigger brains and become the smart species we are today.

Hunting was everything.

"In a very real sense our intellect, interests, emotions and basic social life—all are evolutionary products of the success of the hunting adaptation," wrote leading anthropologists Sherwood Washburn and Chet Lancaster in their chapter of a 1968 book about the symposium, also titled *Man the Hunter*. The importance of the kill, dramatic as it was, would later be popularized for a wider audience in a 1976 book by Robert Ardrey, a Hollywood screenwriter who changed career to focus on anthropology. "It is because we were hunters, because we killed for a living, because we matched wits against the whole of the animal world, that we have the wit to survive even in a world of our own creation," he wrote in *The Hunting Hypothesis*.

But for some anthropologists, this way of characterizing the past struck a bum note. For one thing, it utterly diminished the role of women. This wasn't even a time when sexism could go easily unchecked. Universities were starting to offer courses in women's studies and gender studies, and female life scientists and social scientists were becoming famous in their fields. Primatology was on its way to becoming a female-dominated discipline. How could anthropologists now claim that women were the sidekicks in human history? By the end of the conference a growing cadre of scientists—many of them women, but some men too—were outraged. Already marginalized for decades, the hunting hypothesis was threatening to airbrush women out of the evolutionary story altogether.

Capturing their feelings, in 1970 anthropologist Sally Linton (later publishing under the name Sally Slocum) presented a provocative retort at the American Anthropological Association's annual meeting. It was titled "Woman the Gatherer: Male Bias in Anthropology." Her words echoed those of suffragist Eliza Burt Gamble, whose critique of Charles Darwin and his contemporaries had been published around eighty years earlier. Linton passionately condemned her field as one that had been "developed primarily by white Western males, during a specific period in history." Given this bias, she said, it wasn't surprising that anthropologists had failed to ask just what it was females were doing while the males were out hunting.

"A theory that leaves out half of the human species is unbalanced," Linton announced. "While this reconstruction is certainly ingenious, it gives one the decided impression that only half the species—the male half—did any evolving."

The focus of her complaint was the notion that women were somehow not equal providers for their families. Experts at the 1966 "Man the Hunter" conference already knew this wasn't true. In fact, one organizer, Richard Lee, had been the very anthropologist to establish the immense importance of women in sourcing food. His fieldwork had shown that, while often not hunters of big animals, women were responsible for getting hold of every other kind of food, including plants, roots, and tubers, as well as small animals and fish. Men were the hunters, but women were the gatherers.

Gathering was arguably a more important source of calories than hunting. In 1979 Lee noted that among the !Kung hunter-gatherers in Africa, women's gathering provided as much as two-thirds of food in the group's diet. As well as feeding their families, women were often also responsible for cooking, setting up shelter, and helping with hunts. And they did all this at the same time as being pregnant and raising children.

By elevating hunting, anthropologists were willfully ignoring women, according to Sally Linton. She reasoned that the hunting hypothesis couldn't possibly explain as much about human evolution as it claimed to. If hunting by men was what drove communication, cooperation, and language in our species, then why were there so few psychological differences between men and women? The original social bond in any human society would clearly have been between a mother and her child, she added, not between hunters. And what about the intellectual challenges of raising children? "Caring for a curious, energetic, but still dependent human infant is difficult and demanding. Not only must the infant be watched, it must be taught the customs, dangers and knowledge of its group," she added.

The title of Linton's passionate talk, "Woman the Gatherer," was seen as the female counterpoint to "Man the Hunter." And it became a rallying cry for other researchers who were determined to bring women to the heart of the human evolutionary story.

———

Adrienne Zihlman, now a prominent anthropologist at the University of California, Santa Cruz, had been teaching for a few years by the time Sally Linton addressed the American Anthropological Association in 1970. "It really struck a note," Zihlman tells me. We are sitting in her home in San Francisco, a stack of papers and books in front us on the table. One book, which she wrote a chapter for in 1981, is titled *Woman the Gatherer*.

"Women were invisible. It's hard for you to imagine what that was like. It was making women visible for the first time," Zihlman continues. She was deeply inspired by Linton and decided to follow up on her ideas and build hard data around them, digging up evidence from observations of hunter-gatherers, primates, and fossils. Through detailed research like this, living with hunter-gatherers, and dissecting their lives, anthropologists and ethnologists like her now finally understand just how mobile, active, and hard working women really are.

One important myth to be cracked was that males were always the main inventors and tool users in our past. Zihlman is convinced this is wrong. While chimpanzees tend to pick and eat their food alone and on the spot, at some point in history humans began to gather and bring it back home to share. They would have needed containers to hold all this food, as well as slings to carry their babies while they gathered—and both probably before anyone created stone hunting tools. These are likely to have been the earliest human inventions, she says, and they would have been used by women. One of the earliest tools, meanwhile, would have been the "digging stick." She tells me that female gatherers to this day use digging sticks to uncover roots and tubers and kill small animals. They're as multifunctional as Swiss army knives.

What digging sticks, slings, and food bags all have in common, though, is that they're wooden or made of skin or fiber, which means they break down and disappear over time. They leave no trace in the fossil record, unlike hardwearing stone tools that archaeologists have assumed are used for hunting. This is one reason, adds Zihlman, that women's inventions, and consequently women themselves, may have been neglected by evolutionary researchers.

Other species provide clues, too, that suggest hunting and toolmaking are not exclusively male domains. The primatologist Jane Goodall has shown through her intimate observations of chimpanzees that females are more skilled at using simple tools and cracking nuts with hard shells than males are. This is partly because they spend more time doing it. Zihlman points out in a paper in the journal *Evolutionary Anthropology* in 2012 that young chimpanzees learn from their mothers to "fish" for termites and that their daughters spend more time watching them than their sons. Some chimpanzees have even been spotted hunting for small animals, such as squirrels, using sticks that they bite off into sharp points. "It is predominantly females, particularly adolescent females, that hunt this way, doing so almost three times as often as males," she writes.

Other scientists have also tallied how many calories hunter-gatherers bring home to their families and how this breaks down by sex. They've reinforced earlier observations that the food brought home by women is vital to keeping everyone alive.

Men's contribution to calories from hunting varies hugely, depending both on the society and the environment they're in, explains Richard Bribiescas at Yale University, who has done fieldwork with the !Kung in East Africa and the Aché hunter-gatherers in eastern Paraguay. "For example, in the group that I worked with years ago, the Aché, they were bringing in 60 percent of the calories. In groups like the !Kung, men were bringing in 30 percent of the calories. It also makes a difference in the type of game they're going after. In the !Kung, for example, they were going after very large, high-risk game like giraffe. It was boom or bust. Whereas with the Aché in Paraguay, the largest thing they would hunt would be the tapir, which is about the size of the small pig. They were getting a lot of small animals, which are a lot more reliable. So it really varies with the environment," says Bribiescas.

In a 2002 paper in the *Journal of Human Evolution*, anthropology professors at the University of Utah James O'Connell and Kristen Hawkes confirmed that hunting is rarely a reliable source of food. Observing more than two thousand days of hunting and scavenging, they estimated that the Hadza in northern Tanzania, for instance, successfully brought home a large animal carcass only one hunting day in thirty. In none of the societies that have been studied do men bring home all the food. At worst, they bring in far less than half. This means that relying on male hunting, in many places, would leave families hungry.

"Something beyond family provisioning was needed to explain men's work," Hawkes and her colleagues have written. They've argued that the reason male hunter-gatherers persist with hunting big animals rather than gathering or chasing smaller prey, like women tend to do, is that it offers them an arena to show off to others, boosting their status and attracting mates.

But the question of who does more for the survival of their families remains a bone of contention. Hawkes's observations have been challenged by anthropology professors Michael Gurven at the University of California, Santa Barbara, and Kim Hill at Arizona State University. In a 2009 paper they published in the journal *Current Anthropology*, titled "Why Do Men Hunt?," they revisit the hunting hypothesis. Gathering plants, done mainly by women, can be a risky source of food, they argue. Plants are often seasonal, for instance. And men in some societies, including the Aché hunter-gatherers in Paraguay, do target small, more reliable game, suggesting that they aren't just looking to display their hunting prowess.

Rebecca Bliege Bird, a professor of anthropology at Pennsylvania State University's College of the Liberal Arts, meanwhile believes that researchers such as Gurven and Hill cling to the hunting hypothesis because of the communities they've happened to study, in particular, the Aché. "Some people's ideas about what hunting and gathering were like in the past tend to be shaped by the society they've spent most of their time in," she explains. "In Oceania, Southeast Asia, and sub-Saharan Africa, women contribute a lot to production. And in other places, like South America, women contribute less to production."

She adds that the evidence to date makes the hunting hypothesis nothing less than "old-fashioned and ridiculous."

The other myth around the hunting hypothesis is the question of language and intelligence. Were anthropologists right in thinking that male hunters drove forward the development of human communication and brain size? Sarah Hrdy's work on infants and mothers has supported Sally Linton's suggestion that language probably evolved, not through hunting, but more likely through the complex and subtle interactions between babies and their caregivers. Over generations, Hrdy explains, babies that were just a little better at gauging what others were thinking and feeling were the ones most likely to be cared for. "They have to engage and appeal to others. They have to understand what someone else is going to like," she adds. This quest for engagement could have provided the original urge to communicate, pushing our ancestors beyond simple chimp-like calls toward sophisticated language.

More recent research has bolstered this idea. In the summer of 2016, Steven Piantadosi and Celeste Kidd in the Department of Brain and Cognitive Sciences at the University of Rochester, New York, published evidence in *Proceedings of the National Academy of Sciences* that child care may have been one major factor in driving up human intelligence. Human babies are particularly immature and helpless when they're born, compared to other mammals. One reason for this is that their heads are so big—to make room for their large human brains—that if they were born much later, they simply wouldn't fit through their mothers' birth canals. "Caring for these children, in turn, requires more intelligence—thus even larger brains," write Piantadosi and Kidd.

A runaway evolutionary process, in which brains got even bigger and babies were born even earlier, could explain why humans eventually became as smart as they are now.

———————

The picture all this leaves us with is very different from that of the sedentary, weak, and dependent woman that some evolutionary biologists have painted in the past.

"When you see pictures of what these women can do, they're pretty strong," Adrienne Zihlman tells me. In her chapter in the 1981 book *Woman the Gatherer*, she includes a striking image, shot by anthropologist Richard Lee, showing a seven-month pregnant !Kung woman striding through the Kalahari like an athlete. She's supporting a three-year-old child on her shoulders, brandishing a digging stick in one hand, and hauling the food she's gathered on her back to take home.

Seen from an evolutionary context, strength like this makes sense. Our sedentary lifestyles and beauty ideals that prize skinniness and fragility in women over size and strength can blind us to what women's bodies are capable of. But if the lives of modern-day hunter-gatherers are anything to go by, our female ancestors would have done plenty of hard physical work. Subsistence living, which is the way humans survived for several million years before they settled into food production of their own around ten thousand years ago, is so tough that they wouldn't have had any other choice. Millions of women around the world now still have no option but to do hard, heavy work to survive.

Women are also known to be particularly good at endurance running, notes Marlene Zuk, who runs a lab focusing on evolutionary biology at the University of Minnesota. In her 2013 book *Paleofantasy*, she writes that women's running abilities decline extremely slowly into old age. They've been known to go long distances even while pregnant. One example is Amber Miller, an experienced runner who in 2011 ran the Chicago marathon before giving birth seven hours later. English runner and world record holder Paula Radcliffe has also trained through two pregnancies.

For a large chunk of early human history, when humans migrated out of Africa to the rest of the world, women would have traveled hundreds or thousands of miles, sometimes under extreme environmental conditions. If they were pregnant or carrying infants, the daily physical pressures on them would have been far greater than those faced by men. "Just reproducing and surviving in these conditions, talk about natural selection!" says Zihlman. "Women have to reproduce. That means being pregnant for nine months. They've got to lactate. They've got to carry these kids. There's something about being a human female that was shaped by evolution. There's a lot of mortality along the way that really can account for it."

This may even explain the mystery of why women are on average biologically better survivors than men are. "There is something about the female form, the female psyche, just the whole package, that was honed over thousands and thousands, even millions of years to survive and spread around the world," says Zihlman.

The harsh realities of subsistence living would also have forced women and men to be flexible and share workloads. "The thing about hunter-gatherer societies is that there is less rigid division of human labor because everybody learns everything," she explains. In our ancient past, thousands of years ago, it's even possible that men would have been far more involved in child care and gathering while women would have been hunters.

"Being a woman hunter is a matter of choice."

"I was up the river, and I saw a couple of women with bows and arrows. That was 1972," recounts anthropologist Bion Griffin, an emeritus professor at the University of Hawaii at Manoa. He and fellow anthropologist Agnes Estioko-Griffin (they are married) are speaking to me over an unreliable line from the Philippines, where they both live.

Bion describes his first eye-opening trip to the island of Luzon in the Philippines. It's home to a tiny hunter-gatherer community known as the Nanadukan Agta. Today, logging, farming, and migration have changed the Agta way of life utterly, drawing them away from subsistence living and integrating them into the farms around them. They share this fate with many of the other remaining hunter-gatherers around the world. But forty years ago the Griffins were lucky enough to catch the tail end of the Agta's old way of life. The Nanadukan Agta were then known to fish and hunt regularly for wild game such as pigs and deer, using bows, arrows, and the help of dogs.

What made them unusual, though, was that Agta women hunted and fished.

Women hunters are not unheard of. In the 1970s the scientific literature included a few references to female hunters scattered across the globe, all the way from the Tiwi people in hot Australia to the Inuit in the cold Arctic north. But Nanadukan Agta women were perhaps the most enthusiastic and regular female hunters of all. "We found first of all that within this particular group, a considerable number of women hunted," Bion tells me. "A lot of women don't carry bows but will use knives, or knives strapped onto a sapling that's been cut down, in order to finish off a cornered deer or a pig that the dogs are holding. . . . And we found that there were a few women that loved to hunt. We found out that they were very successful in hunting."

Women hunted even when they had alternative ways of feeding themselves, adds Agnes. She recounts one time when the men of the group went off for several days on a hunt. Rather than gathering roots or fruits or trading with local farmers, a group of women went out on their own and killed a pig. "It was their choice to go off hunting," she explains. Bion adds, "It varies from women opportunistically hunting and killing when traveling in the forest, including when they're carrying their babies and kids with them, to young grandmothers or very mature women who had a long history of hunting but who have no real demands on child care, except the usual grandmothers always helping out and taking care."

Agnes Estioko-Griffin published some of these findings in a paper in 1985. She noted that every able-bodied Agta, male or female, knew how to spearfish. Of twenty-one women above the age of fourteen in the group, fifteen were hunters, four had hunted in the past, and only two didn't know how to hunt. In half of all the hunting trips she observed, men and women hunted together. If there were differences, they were in the way women tended to hunt. For instance, a woman never went alone, to avoid the risk of people suspecting that she was having a secret tryst with a lover. Women hunters were also more likely to use dogs to help with the kill.

"Being a woman hunter is a matter of choice. To keep an individual from performing certain tasks due to biological reasons is unthinkable to the Agta," she described. "Lactation may temporarily cause a decrease in a woman's active participation in hunting, but it certainly does not preclude her involvement in this activity."

The key to making this possible was cooperative breeding, she adds. Women would take nursing infants with them on the hunt and leave older children in the care of other family members back home. Or a woman might nurse her sister's baby while she was out hunting. "Even young adults could do the babysitting or keep an eye on the smaller children, cousins or siblings, left behind at the camp. Cooperative breeding is, I think, a very important component," she explains.

The more the couple explored, the more they found that the women and men of the Nanadukan Agta were able and expected to do the same jobs. "By and large, people did whatever they wanted to do," says Bion Griffin. No sphere of work was exclusively male or female, except perhaps killing other people. Women would stay back when groups of men went out on warlike enemy raids. "Some men did all sorts of child care, cooking, and so on. Others didn't bother much with, say, cooking. I think everybody did everything. The only thing I can recall, I don't recall men ever weaving baskets. But then, no one weaves baskets much. Men built houses with the women, men attended babies, they gathered firewood, they cooked, they pounded rice when there was rice to pound."

Even as their old way of life disappears, the Nanadukan Agta have shown that, beyond the biological fact that women give birth and lactate, culture can dictate almost every aspect of what women and men do. The way lives are divided when it comes to child care, cooking, getting food, hunting, and other work is a moveable feast. There's no biological commandment that says women are natural homemakers and unnatural hunters or that hands-on fathers are breaking some eternal code of the sexes.

The dilemma they pose for evolutionary biologists, though, is why they are the exception rather than the rule. Why don't women hunter-gatherers everywhere hunt? And why aren't all human societies just as egalitarian?

———————

We sometimes imagine sexual equality to be a modern invention, a product of our enlightened, liberal societies. In actual fact, anthropologists have long known that the way women are treated throughout the world wasn't always like this.

Anthropologist Mark Dyble, based at University College London, has studied another Agta community in the Philippines, known as the Palanan Agta, and analyzed this data together with that from a more distant group of hunter-gatherers in the Congo, a subgroup of the BaYaka, known as the Mbendjele. His research reveals a connection between the social structure of hunter-gatherer communities and high levels of sexual equality. It's evidence, he suggests, that equality was a feature of early human society before the advent of agriculture and farming.

Published in 2015 in the journal *Science*, Dyble's work built up detailed genealogies of hundreds of adults in these two communities. "We know as much as they know about their family histories. We even know if someone is second cousin with someone," he tells me. These genealogies reveal that people living together are generally unrelated to each other. Women don't always live with or near their husbands' families and the same is true of men and their wives' families. Sometimes they will switch between families, and sometimes they won't live with close family at all.

Given the choice, people usually prefer to live with their own relatives all the time, because of the support and protection they can give them. "It's not that individuals don't want to live with kin," Dyble explains. "It's just that if everyone tried to live with as many kin as possible, this places a constraint on how closely related communities can be." And this in turn means that neither men nor women have greater control over whom they live with. There must be sexual equality in decision making. "It has this transformative effect on social organization," he says.

If this arrangement was normal in our evolutionary history, Dyble believes it could explain some aspects of human development. "We have the ability to cooperate with unrelated individuals, which is different from what we see in primates, which are very wary of interacting with individuals they haven't met before," he says. This is crucial to complex society. If people couldn't cooperate with people they weren't related to, civilization as we know it simply couldn't exist. A study by anthropologist Kim Hill and his colleagues, published in the journal *PLOS ONE* in 2014, confirms that hunter-gatherers do interact widely with others. Their own data from the Aché in eastern Paraguay and Hadza in Tanzania suggest a person's social universe can include as many as a thousand people over a lifetime. A male chimpanzee, by contrast, will only ever interact with around twenty other males.

This all points to the possibility that the way the Palanan Agta used to live may have been usual in our past. Historical investigations have always failed to uncover good evidence for matriarchal societies, in which women hold the reins of power. But that doesn't mean humans weren't egalitarian.

"There's a general consensus now that hunting-gathering societies, while not perfectly egalitarian, were less unequal, particularly with regard to gender equality," agrees Melvin Konner, a professor of anthropology at Emory University in Atlanta, who has spent many years doing fieldwork with hunter-gatherers in Africa. The communities he has studied have very little specialization of roles, he explains. There are no merchants or priests or government. "Because of the scale of the group dynamics, it would be impossible for men to exclude women. . . . Men and women participated, if not equally, women contributed at least 30 to 40 percent of the time."

If the women of the Nanadukan Agta persisted with hunting for so long while others abandoned it earlier, one reason might have been their environment. The tropical forests in Luzon have fewer large and dangerous animals than in other parts of the world, such as South America, says Bion Griffin. Michael Gurven and Kim Hill, who have catalogued the reasons women don't hunt, suggest that women avoid hunting as the risk of death rises. This is important to a group's overall survival, because losing a mother is far more dangerous for a child than losing a father. In some societies and environments, hunting isn't just dangerous; it can also take women far away from their home base for days at a time. If the culture does not provide enough support for women in terms of child care or other work, a woman may simply be unable to put in as many hours as a man to perfect her skills, making her a less useful killer.

Bion Griffin tells me that much of the resistance to the idea of women hunters comes from evolutionary theorists who can't accept that hunting and motherhood are compatible. But among the Agta, hunting didn't seem to put children at greater risk, as far as he and Agnes Estioko-Griffin could tell. It only brought in more food for everyone in a community in which food would otherwise have been desperately scarce.

Anthropologist Rebecca Bliege Bird, who has studied women hunter-gatherers in Australia, agrees. "There's no reason why women wouldn't hunt where hunting is an economically productive and predictable thing to do," she says. One example she gives is that of the Meriam, an indigenous Australian society living in the Torres Strait Islands. They are skilled seafarers. On the beach, men spend more time line fishing, in the hope of bringing home a large, prized catch, while women choose to go after resident reef and shellfish where the odds of success are higher. As a result, women's fishing harvests are more consistent and sometimes even more productive than men's. "In most circumstances, hunting of large animals is not a very productive thing to do. I would guess that the majority of subsistence for most hunter-gatherers in most environments is the small animals. And women are going to be the major procurers of small animals," she says.

Another example from the same continent is the Martu, an aboriginal tribe in Western Australia for whom hunting is a sport. Outrunning animals is a skill perfected by women in particular. "When Martu women hunt, one of their favorite prey are feral cats. It's not a very productive activity, but it's a chance for women to show off their skill acquisition. Women gain huge notoriety going after these cats," Bliege Bird tells me. The hunting is done in scorching summer heat. "Women chase after these cats. They run to tire them out. It's just tremendous the amount of effort that goes into it."

Even among the Aché in eastern Paraguay—a community in which women don't hunt—there is evidence that women are still able to hunt if they want to. Ana Magdalena Hurtado, an evolutionary anthropologist at Arizona State University, has documented how Aché women act as "eyes and ears" for male hunters. She and her colleagues once saw an Aché woman hunting while carrying an infant. They concluded, "Aché women are capable of hunting but avoid doing so most of the time." Their focus, instead, must be on other work.

When it comes to family and working life, the biological rule seems to be that there were never any rules. While the realities of childbirth and lactation are fixed, culture and environment can dictate how women live just as much as their bodies do.

For those who have spent their careers on the outside looking in, documenting these rare human societies whose ways challenge our stereotypes, this can be personally life changing. At the end of our interview, Bion Griffin and Agnes Estioko-Griffin tell me that there's no sexual division of labor in their own household, just like there was none among the Nanadukan Agta they studied for so many years. "And so, I'm off to cook dinner now!" Bion laughs before he hangs up the phone.

At home in London, I realize with disappointment, I'm the one cooking dinner that night.

CHAPTER 6

Choosy, Not Chaste

If the world were ours too, if we believed we could get away with it, . . . the force of female desire would be so great that society would truly have to reckon with what women want, in bed and in the world.

—Naomi Wolf, *The Beauty Myth*, 1990

You're at university and a stranger of the opposite sex sidles up to you. "I've been noticing you around campus. I find you to be very attractive," they say. Before you know it, the mysterious person is inviting you back to their room to sleep with them.

It may be the least creative way of picking someone up, but if it works on you, then research suggests you're almost certainly a man. This scenario was part of a real experiment at Florida State University conducted in 1978 and designed by psychology professors Russell Clark and Elaine Hatfield to settle a classroom dispute over whether, compared to women, men are more open to casual sex. Their method was simple. They recruited a bunch of young volunteers from an experimental psychology class, none of them too bad looking but none wildly attractive either, to approach people across campus and repeat the same pickup line. This was followed by one of three requests: to go out on a date, to go to their apartment, or to go to bed with them.

The results were stark. Even though men and women were equally likely to go on a date with a stranger, none of the women would sleep with one. Three-quarters of the men, on the other hand, were willing to have sex with a woman they didn't know. When the psychologists repeated the experiment in 1982, the results were almost the same. The women, they observed, were often appalled at being propositioned in this way. "What is wrong with you?

Leave me alone," one said. The men were a different story, even apologizing when they refused. "In fact they were less willing to accept an invitation to date than to have sexual relations!" Clark and Hatfield noted.

For years they struggled to get their paper published for fear on the part of publishers that it was too frivolous. When it finally came out in 1989 in the *Journal of Psychology and Human Sexuality* under the title "Gender Differences in Receptivity to Sexual Offers," it became a classic. After all, it neatly confirmed what everyone thought they already knew about sex and the sexes. Men are naturally polygamous and just fighting nature when they become tied into long-term relationships. Women are monogamous and always looking for the perfect partner.

It comes down to the fact, some biologists say, that males and females want fundamentally different things. They're stuck in an endless evolutionary tussle—one indiscriminately chasing any female to boost his odds of fathering the most children, and the other trying to escape unwanted male attention in the careful search for the best-quality father for her offspring. Charles Darwin himself had laid this observation in scientific stone back in 1871 in his famous work *The Descent of Man, and Selection in Relation to Sex*.

The idea was even experimentally tested in 1948 in another mating experiment. This one wasn't on humans, though, but on a humble little fly that appears when fruit rots.

———————

When it comes to reproduction, the easiest species to study are those that mate quickly and breed abundantly. Humans are not that species.

Angus John Bateman, a botanist and geneticist working at the John Innes Horticultural Institute in London in 1948, was wise enough to pick the common fruit fly, a creature that lives so hard and fast that it's sexually mature within a few days of birth and can lay hundreds of eggs at a time. What makes the fruit fly a scientist's best friend is that it has genetic mutations that make each one look slightly different from the next, depending on what it inherits, such as curlier wings or narrower eyes. By tracking these differences, Bateman could reliably pick out which fly belonged to which parents. From this, he knew which flies were mating successfully.

Like Hatfield and Clark's experiment, Bateman's was simple. He took three to five adult females and the same number of adult males, then watched to see how they performed in the mating game. A fifth of the male flies, he found, didn't manage to produce any offspring, compared with only 4 percent of the females. The most successful male flies, though, produced nearly three times as many offspring as the most successful female fly. None of the females were short of offers, but the least successful males suffered routine rejection. It confirmed Darwin's long-standing theory that males in species like these are more promiscuous and less discriminating, while females are pickier and more chaste.

"Darwin took it as a matter of general observation that males were eager to pair with any female, whereas the female, though passive, exerted choice," wrote Bateman. The fruit fly species he studied "seems to be no exception to the rule."

Darwin had reasoned that when one sex has to compete for mates, there's greater pressure on it to evolve the features the other sex is looking for. It needs to be strong enough to beat off the competition, too. He called this evolutionary process "sexual selection." And his observations suggested that males faced far more of this pressure than females. This would explain why the males of certain species, including our own, tend to be bigger and stronger than the females. It explains, too, such marvels of nature as the lion's giant mane and the peacock's flamboyant blue and green plumage. There don't seem to be any reasons why lions need manes or peacocks need such cumbersome, fancy feathers except to attract the opposite sex.

"There is nearly always a combination of an undiscriminating eagerness in the males and a discriminating passivity in the females," wrote Bateman. His fruit fly experiment reinforced Darwin's theory that sexual selection acts more heavily on males than on females. Some male flies were studs, others were duds, but none of them for want of trying. The competition was intense enough that a few did far better than the rest. The female flies, meanwhile, seemed to be comfortable in the knowledge that they could choose the males they wanted. They seemed to be under little pressure at all. In fact, according to Bateman, a tiny number were even willing to forgo mating for the moment if, presumably, they didn't see what they liked.

Bateman's observations of fruit flies, extrapolated to other species including our own, would renew scientific interest in sexual selection theory. But not immediately. His paper lay beneath the radar for decades. He never wrote about sexual selection again. It wasn't until twenty-four years later that his fruit fly experiment was finally popularized by a young researcher called Robert Trivers.

———————

Trivers, age seventy-three, has had a colorful life for a biologist.

His website, which promotes his autobiography—appropriately titled *Wild Life*—says that he's spent time behind bars, that he founded an armed group to protect gay men in Jamaica from violence, and that he once drove a getaway car for a founder of the Black Panthers, the black nationalist organization active in the sixties and seventies. He was also the biologist who once told a reporter that biologist Sarah Blaffer Hrdy should focus on being a mother rather than on her career.

Today Trivers lives on a rural estate he's bought in Jamaica. When I interview him over the phone, he tells me that he and the workers there call it "Man Town," because there are no women around. When I ask him where he works nowadays, he says that he's in a dispute with his employer, Rutgers University in New Jersey, which means that soon he'll be out of a job. Apparently, he had been forced to teach classes on subjects he didn't know anything about.

However much of a roller coaster his life has been, Trivers is considered one of the most influential evolutionary biologists in the world, in particular for theories he developed early in his career. A paper he published in 1972 about Angus Bateman's 1948 fruit fly experiment has been cited by researchers at least eleven thousand times. Titled "Parental Investment and Sexual Selection," it has fundamentally shaped the way researchers today understand sexual selection.

Trivers was just a young researcher at Harvard University, studying mating pigeons outside his window, when one of his tutors suggested he look up Bateman's work. And he remembers it with graphic clarity. He went to the museum to photocopy it, "with my testicles firmly pressed against the side of the Xerox machine," he tells me, with a throaty laugh. As soon as he read it, "The scales fell from my eyes," he says. It would mark a turning point in his career.

He realized that females must be choosier and less promiscuous than males because they have a lot more to lose as parents from making a bad choice. Take the example of humans: men produce lots of sperm and don't necessarily need to invest in their children, while women have only a couple of eggs to fertilize at a time, followed by nine months of pregnancy and many years of breast-feeding and child raising. "The logic was obvious after a moment's reflection. You know the female is spending a lot producing those two eggs, and the male is spending a day's ejaculate, which is trivial," he explains. "When I lecture to students I sometimes point out that, during the last hour, every testicle in the room has generated a hundred million sperm. That's a lot of sperm with nowhere to go."

In his 1972 paper about Angus Bateman's observations of fruit flies, Trivers writes, "A female's reproductive success did not increase much, if any, after the first copulation and not at all after the second." A female, he suggests, gains nothing from adding extra notches to her belt. One male is enough to get her pregnant, and once pregnant, she can't be any more pregnant. "Most females were uninterested in copulating more than once or twice."

This theory implies that when parental investment changes, so might sexual behavior. In monogamous species in which fathers are much more heavily involved in child care, these rules could theoretically reverse. The more that males invest time and energy in their children, the choosier they might become about whom they mate with and the more competitive females might become for their attention. And, indeed, in certain monogamous species of bird, it's the females that chase after the males.

In humans, of course, many men are reliable fathers who invest as much as mothers in raising children. But Bateman didn't believe this would necessarily change how men behave. He wrote that even in monogamous species with fairly equal numbers of males and females, the old pattern of sexual behavior—undiscriminating eager males and discriminating passive females—"might be expected to persist as a relic." In his own paper, twenty-four years after Bateman's, Trivers suggests, "In species where there has been strong selection for male parental care, it is more likely that a mixed strategy will be the optimal male course—to help a single female raise young, while not passing up opportunities to mate with other females whom he will not aid."

In other words, he's saying that men are unlikely to have escaped the evolutionary urge to cheat.

"Sounding sexist is not a good reason to ban a theory."

The August 1978 issue of *Playboy* magazine carried a sensational story. "Do Men *Need* to Cheat on Their Women? A New Science Says Yes," boasted the cover. The photograph next to the provocative headline coincidentally featured a model in white suspenders and strappy heels for an item on sexy secretaries. Her pad and pen were carelessly tossed to the floor while she stood pressed against her boss.

The publication of Robert Trivers's paper marked a watershed not only in the way scientists understood sexual behavior but also in how the everyday woman and man in the street understood it. Sexual selection theory, revamped for the twentieth century, rapidly became a tool to explain women's and men's relationship habits. Bateman's theories, once almost forgotten, were transformed into a fully blown set of universal principles, cited hundreds of times and considered solid as a rock. On that rock now rests an entire field of work on sex differences.

In 1979 prominent anthropologist Don Symons, now an emeritus professor at the University of California, Santa Barbara, in his seminal book *The Evolution of Human Sexuality*, reinforced the idea that men seek out sexual novelty while women look for stable, monogamous relationships. "The enormous sex differences in minimum parental investment and in reproductive opportunities and constraints explain why *Homo sapiens*, a species with only moderate sex differences in structure, exhibits profound sex differences in psyche," he writes. One of Symons's theories is that the female orgasm isn't an evolutionary adaptation but a by-product of the male orgasm, just like male nipples are a vestige of female nipples. If women do experience orgasm, he implies that it's only a happy biological accident.

An unimpressed critic at the time, Clifford Geertz at the Institute of Advanced Study, Princeton, summed up Symons's book in the *New York Review of Books* with the old verse, "Higgamous, Hoggamous, woman's monogamous; Hoggamous, Higgamous, man is polygamous."

Despite the skepticism, within a couple of decades of his book being published, the science had gone mainstream. Robert Trivers's work, by drawing human behavior further into the realm of evolutionary biology, had helped spawn an entire field of research known today as evolutionary psychology. One of the world's most well-known academics in this subject is David Buss, who now teaches at the University of Texas at Austin. In his book *The Evolution of Desire: Strategies of Human Mating*, published in 1994, he writes, "Because men's and women's desires differ, the qualities they must display differ," adding that it makes sense for women to be naturally monogamous because "women over evolutionary history could often garner far more resources for their children through a single spouse than through several temporary sex partners."

This idea popped up again in a 1998 *New Yorker* article by the cognitive psychologist Steven Pinker. Under the title "Boys Will Be Boys," he used evolutionary psychology to defend US president Bill Clinton, whose affair with his intern Monica Lewinsky had just been made public. "Most human drives have ancient Darwinian rationales," he writes. "A prehistoric man who slept with fifty women could have sired fifty children, and would have been more likely to have descendants who shared his tastes. A woman who slept with fifty men would have no more descendants than a woman who slept with one." Pinker has described Don Symons's book as "groundbreaking" and Robert Trivers's work as "monumental." He was also among those who stood up for Harvard University president Lawrence Summers when he suggested that innate sex differences might explain the shortfall of top female scientists.

The scope of Charles Darwin's original work on sexual selection stretched far beyond sexual behavior, of course. It wasn't just about mating habits but also about how the pressure to attract the opposite sex would have acted more heavily on males, influencing their evolutionary development by forcing them to become more attractive and smart. In the *Descent of Man* in 1871, he wrote, "The chief distinction in the intellectual powers of the two sexes is shewn by man attaining to a higher eminence, in whatever he takes up, than woman can attain. . . . Thus man has ultimately become superior to woman."

More than a century later, even this controversial aspect of sexual selection theory has been resurrected. In 2000, an evolutionary psychologist at the University of New Mexico, Geoffrey Miller, published *The Mating Mind: How Sexual Choice Shaped the Evolution of Human Nature*, in search of what he calls "a theory for human mental evolution." Females in our evolutionary past may have developed a preference for males who were better at singing or talking, he writes. As men became more creative and intelligent and better at singing and talking, they would have become more attractive and successful at mating. Through a "runaway process" in which smarter males mated more often and sired smarter offspring, Miller argues, the human brain could have reached its relatively large size as quickly as it did.

"Male nightingales sing more and male peacocks display more impressive visual ornaments. Male humans sing and talk more in public gatherings, and produce more paintings and architecture," he writes. Later he adds, "Men write more books. Men give more lectures. Men ask more questions after lectures. Men dominate mixed-sex committee discussions." Men are better at all these things, he implies, because they have evolved to be better.

For anyone who fears this might be a little unfair to women, Miller has a response. "In the game of science," he advises his readers, "sounding sexist is not a good reason to ban a theory."

"Multiple mating is very, very common among females."

At the heart of sexual selection theory, as it applies to humans at least, is the notion that men are promiscuous and undiscriminating while women are highly discriminating and sexually passive. Females are choosy and chaste. It all comes down to Angus Bateman's principles, as demonstrated both by his flies and by Clark and Hatfield on the campus of Florida State University in 1978. Men will sleep with strangers while women simply won't.

But not everyone is convinced this is true.

Today there is a huge body of research that flies in the face of Bateman's principles. It has been building up for many decades. Anthropologist and primatologist Sarah Hrdy's research on the Hanuman langurs of Mount Abu forty years ago showed that a female monkey can benefit from mating with more than one male because it confuses them all over their possible paternity of her children, making them less likely to commit infanticide. In her vivid studies of red colobus monkeys in the Abuko Nature Reserve in Gambia, London-based anthropologist Dawn Starin also describes how sexually confident female primates can be. "When it came to sex, she was nothing if not assertive," she writes in a 2008 issue of *Africa Geographic*, about a monkey she saw. "For a few months every year, the forest is taken over by a bunch of female hooligans, strutting their stuff, giving guys the eye and luring nervous males into the bushes."

In more distant species from us, researchers have found similar evidence of females mating with multiple males. Many birds thought to be monogamous have turned out not to be. Female bluebirds have been spotted flying considerable distances at night just to mate with other males. Data on the small-mouthed salamander, bush crickets, yellow-pine chipmunks, prairie dogs, and mealworm beetles have shown that the females of all these species, too, enjoy more reproductive success when they mate with more males.

"It's pretty widespread. Some would even say ubiquitous. Multiple mating is very, very common among females," says animal behaviorist Zuleyma Tang-Martínez from the University of Missouri, Saint Louis. She tells me that as a graduate student she was as convinced of Bateman's logic as anyone. "It's a very simple idea. It makes sense in terms of the cultural stereotypes we have, and so you buy into it," she says. "It was only when I sort of matured as a scientist that I started asking questions, and I started seeing evidence come out that didn't go along with Bateman, that I started to take a much more thorough look at the evidence."

Tang-Martínez has spent years dissecting the facts around Bateman's principles and published numerous papers on his ideas. Her conclusion is that the sheer weight of evidence should be enough to force scientists to rethink Bateman's principles. In fact, she adds, a paradigm shift is already underway. Scientific understanding around the breadth of female sexual nature has expanded to better encompass the true variety in the animal kingdom. Far from being passive, coy, and monogamous, females of many species have been shown to be active, powerful, and very willing to mate with more than one male.

However, the shift has been slow to come in part because of huge amounts of resistance along the way. In his 1982 review of Sarah Hrdy's book *The Woman That Never Evolved*—which presents more evidence contradicting the image of the coy, chaste female—anthropologist Don Symons raised his eyebrows, especially at her suggestion that, like the female langurs at Mount Abu, evolution might favor females that are sexually assertive and competitive. "In promoting her view of women's sexual nature, Hrdy provides dubious evidence that this nature exists," Symons wrote, dismissively.

According to Sarah Hrdy, this hostility toward viewpoints like hers hasn't gone away. "It is impossible to understand this history without taking into account the background, including the gender, of the researchers involved," she wrote in a chapter of *Feminist Approaches to Science*, published in 1986. In her own review of Don Symons's book on human sexuality from 1979, she referred to this old-fashioned way of thinking as "a gentlemanly breeze from the nineteenth century." She believes that, just like in Darwin's time, scientists have twisted sexual selection theory in ways that are unfair not only to women but also to the truth.

"Sexual selection is brilliantly insightful. Darwin got that exactly right. The problem was that it was too narrow and it didn't explain everything," Hrdy tells me. Some of the most powerful evidence against Bateman's principles isn't even in other species but in our own, adds Zuleyma Tang-Martínez. "If there's any place that I think I would be extremely reluctant, to put it mildly, to say that Bateman applies, it would be humans," she warns. "I think it's a huge mistake."

———

"Around half of societies say female infidelity is either common or very common," says Brooke Scelza, a human behavioral ecologist at the University of California, Los Angeles. She has a playpen in the corner of her office, and as a young working mother myself I immediately empathize with her.

It's Scelza's empathy with women, in turn, that has given her a unique insight into the cultures she has studied around the world. They include the Himba, an indigenous society of partly nomadic livestock farmers living in northern Namibia. The reason the Himba are vital to understanding the true breadth of female sexuality is because on the spectrum of sexual freedom, Himba women are at a far end. Their culture has a relaxed attitude to women having affairs with other men while they're married, offering them more autonomy and choice over who they have sex with than women in almost any other part of the world.

Carrying out interviews about their marital history, Scelza found that Himba women would tell her which children were fathered by their husbands, but then use the local word *omoka* to describe their other children. "It means you get your water from someplace else. So it's a euphemism. Basically, it's a word they use to describe a child that's either born out of wedlock or who is born through an affair," explains Scelza. Husbands, too, would admit quite openly which of their wives' children they thought were their own and which they thought were someone else's.

Although there's no reason to think men and women don't feel jealous, adds Scelza, the cultural norm among the Himba is that it's as acceptable for women to have affairs as it is for men, and husbands simply have to accept them. They profoundly challenge Angus Bateman's theory that women aren't eager for sex or that they don't want more than one sexual partner at a time.

When Scelza started doing fieldwork with the Himba in 2010, women would ask her why she didn't have men coming to her hut. "Well, I said, 'You know, I'm married.' And they said, 'Yeah, yeah, but that doesn't matter. He's not here.' So then I tried to explain that my marriage was a love match, because then I thought they would understand. And they said, 'It doesn't matter. It's okay, it's okay. He's not going to know; it's okay,'" she recalls. "They really hold a very different idea in their heads about love and sex, that it wouldn't be a bad thing at all for me to say, on the one hand, that I really love my husband but that I'll still be having sex with somebody else when we're apart. That, to them, was not a transgression."

In his 1972 paper on Bateman's fruit fly experiment, biologist Robert Trivers had said that this behavior could have no evolutionary benefits for females. One man is enough to get a woman pregnant, and this marks the limit of her reproductive capacity. More lovers can't make her any more successful at having children. But Scelza has found that statistically this isn't true. "It turned out that having some kids through affairs was actually good for your overall reproduction," she explains.

She's still in the process of collecting data and figuring out the reasons for this. It may be no more than a random correlation, perhaps because the most fertile and highest quality women, who would have more children anyway, attract the most partners. Another factor, of course, is that not every man is as fertile or as good a father as the next. But she adds that there are other reasons why births and child survival go up as women mate with more men. Economics, for example; they might bring in more resources or protection.

Another one is sexual compatibility. Among the Himba, arranged marriages are common, which means women don't always get the husband of their choice. Affairs offer them a workaround by giving them the benefit of a committed, reliable husband at home, as well as the man or men they are more sexually compatible with, away from home.

There's some early research indicating, in other species at least, that when a female chooses the male she wants, her offspring are more likely to survive. In 1999, at the annual meeting of the Animal Behavior Society, Patricia Gowaty, who was then at the University of Georgia, and Cynthia Bluhm at the Delta Waterfowl and Wetlands Research Station in Manitoba, Canada, reported this effect in female mallard ducks. Mallards form pair bonds, but male ducks often viciously harass females into mating with them. When a female duck was allowed to choose her mate free from harassment, her ducklings survived better, Gowaty and Bluhm told *Science News*. Gowaty, working with another team, has seen similar results in house mice.

The Himba, however, are just one band in the rainbow of human behavior. Himba women have the sexual freedom they do partly because of the unusual way in which their society is organized. Women keep close ties to their mothers and childhood families after they get married, which makes it easier for them to leave their husbands and do what they want without disapproval or control. Also, wealth isn't passed down from a father to his children but from a brother to his brother or to his sister's sons, which means that a man may be less concerned with knowing his children are his own. Whoever inherits his cows is guaranteed to be a genetic relative.

In a 2013 paper published in the journal *Evolutionary Anthropology*, "Choosy But Not Chaste: Multiple Mating in Human Females," Brooke Scelza lists a few other places in which women have more than one partner. The Mosuo of China, one of the few societies in the world in which women

head households and property is passed down the female line, people practice what is known as "walking marriage." This allows a woman to have as
many sexual partners as she likes. The lover of her choice simply comes to
her room at night and leaves the next morning. What marks the Mosuo
apart is that men traditionally don't provide much economic or social support to their children.

Similarly, in other small-scale societies where women contribute more
to the family plate, women tend to have more sexual freedom. In the United
States, notes Scelza, "in sub-populations in which reliability on male resources is low as a result of high incarceration rates and unemployment,
female kin provide critical instrumental and emotional support, and patterns
of serial monogamy are common."

Another example is in South America, where some isolated societies
practice partible paternity, the belief that more than one man can be the
father of a baby. In a paper on the topic in the journal *Proceedings of the
National Academy of Sciences* in 2010, Robert Walker and Mark Flinn at the
University of Missouri, Columbia, and Kim Hill at Arizona State University
write, "On the universal partible paternity end of the spectrum, nearly all
offspring have purported multiple cofathers, extramarital relations are normal, and sexual joking is commonplace."

Tracking reproductive success in populations across the world, including Finland, Iran, Brazil, and Mali, researchers Gillian Brown and Kevin
Laland at the University of St. Andrews and Monique Borgerhoff Mulder
at the University of California, Davis, similarly found huge variation. In
their paper published in the journal *Trends in Ecology and Evolution* in 2009
they said the data are "inconsistent with the universal sex roles that Bateman
envisaged."

All this, says Scelza, punctures the biological model of the coy, chaste
female. Working with the Himba, who have a sexual culture so different
from her own, has taught her that the rules about how women and men
behave in relationships have far more to do with society than biology. The
Himba aren't a breed apart. They're just culturally different. "It's not that
they don't have love. It's not that sex has replaced love in this society. They
feel jealous. But the cultural norms that are in place prevent men from really
being able to act upon it," she explains. "If he was, for example, to hit his
wife or something like that, which in some places in the world is completely
an acceptable response, there would be a backlash. He would probably end
up having to pay a fine and be punished for that action."

If there is a difference in sexual behavior, adds Scelza, it's that Himba women seem to be more discriminating than the men. "I think they're still being picky. But I think being picky doesn't mean one partner and you have to stick with them for life."

_____

Where does this all leave Angus Bateman's cherished principles?

As more evidence rolls in, researchers have started to further question the scientific orthodoxy that females are generally more passive and chaste than males. Even the famous 1978 experiment on the campus of Florida State University—which found that men were overwhelmingly more open than women to casual sex with strangers—has been repeated, with surprising results.

"I felt like it wasn't telling the whole story," explains psychologist Andreas Baranowski from Johannes Gutenberg University. In the summer of 2013, he and colleague Heiko Hecht decided to run Clark and Hatfield's seminal study again, this time controlling for certain factors they felt might have affected the original outcome. They were driven by their own personal observations of dating and sex. They instinctively didn't believe that the Florida State University experiment had captured the true spectrum of how women behave. "It wasn't what my experience was in Germany here, or in Europe in general. And also of other colleagues and friends," Baranowski tells me. "My female friends would tell me about hookups and stories about how they would engage in sexual relationships with men, and that's also not represented in the data at all. So it was a bit like, that is weird."

Baranowski and Hecht suspected that women might reasonably be put off having sex with a stranger for lots of good reasons, including the social stigma of getting picked up so casually and, more obvious, the risk that they might be attacked. "We wanted to find out how the original findings would stand up to a more naturalistic setting, such as a cocktail bar, and a more safe setting, namely a laboratory," they wrote in their paper, published in the journal *Archives of Sexual Behavior* in 2015. They wanted to make sure they didn't veer too far from the original experiment either, so they ran it on a university campus as well.

Both on the campus and in the cocktail bar, they got fairly similar results to Clark and Hatfield, with slightly more men than women agreeing to a date, and many more men agreeing to sex. In both cases, though, men weren't nearly as keen to go on dates or have sex compared to the Florida State University experiment. It wasn't proof that Clark and Hatfield had got it wrong, but it was certainly evidence that different places and times can yield different results.

And this was crucial in showing that there's no one way in which the sexes typically behave. The original experiment just wasn't representative. "It's really one dimensional, representing the dating market in the United States on a university campus in the seventies. That's how I felt about it," says Baranowski. "I didn't doubt that they did proper protocol. I think they did. It's just a microcosm there, where they did the experiment."

Where their data got really interesting, though, was in the lab. They wanted their subjects to believe they were going on genuine dates with real people, so the researchers concocted an elaborate ruse based on a dating study. Each person was shown ten photographs of strangers of the opposite sex and told that all these strangers wanted to go on a date or meet up for sex with the subject in particular. If they agreed to meet, they were given a safe environment, and Baranowski and Hecht's research team would then film the first half of their encounter.

All the men in the study agreed to go on a date and also have sex with at least one of the women in the photographs. For women, the figure was 97 percent agreeing to a date and, unlike the first experiment, "almost all women agreed to have sex," says Baranowski.

It was evidence, they noted in their paper, that gender differences are significantly smaller in a nonthreatening environment. It may not have been biology holding women back in the Florida State University experiment but other reasons, most likely social and cultural—like the fear of violence or a moral double standard. One sex difference they did notice in the laboratory setting, though, was that women tended to pick out fewer partners from the photographs they were offered. Like Brooke Scelza found with the Himba in Namibia, they were choosier than the men, but not less chaste.

"We can't just go on pretending that everything is hunky-dory."

"Things like Bateman's principles actually don't make sense to me," says Patricia Gowaty, distinguished professor at the University of California, Los Angeles.

We're sitting on the patio of her home on a mountain in Topanga, nestled in a sprawling state park in Los Angeles County. We're surrounded by wildlife. At one point during our meeting, a wild deer wanders nearby. Gowaty is an animal expert, an evolutionary biologist and a firebrand who has spent her career, which spans five decades, to leaching sexism out of her field by challenging its basic assumptions. Her most famous target has been Angus Bateman's 1948 experiment showing that male fruit flies are more promiscuous than females.

"I became a scientist at the same time as I was becoming a feminist. They were coincident," she tells me. Gowaty's feminism has never waned. It influences her now as much as it did when she had her first job in the education department at the Bronx Zoo in New York in 1967. "In the late 1960s, all over the country, there were groups that were coming together to consciousness raise. The idea of consciousness-raising was simply to talk and to bring to consciousness the ideas associated with the feminism that was emerging at that time." Through forums like these, she began to understand how women throughout history, including her own mother, had been constrained. Their achievements were against the odds.

"There are many women of my generation who have published with their initials to hide their gender," she tells me.

Gowaty was angered, just as her contemporaries Sarah Hrdy and Adrienne Zihlman were, by how evolutionary biology was ignoring and misunderstanding women. Bateman's principles lay beneath some of the claims that angered her most. She spent thirty years studying the mating behavior of Eastern bluebirds, and in the 1970s, when she suggested that female birds were flying away to mate with males that weren't their partners, she simply wasn't believed. Her male colleagues couldn't accept it. They told her instead that the female bluebirds must have been raped.

"I think one of the things that Bateman's principles do is they obfuscate variation in females. So suddenly, there's nothing interesting about females. That's one of the things that bothers me about it. There's embedded sexism there, I think," she says. "They may as well be tenets of the faith."

Gowaty knew that the ultimate test of any scientific experiment rests on the ability to replicate it. So in the 1990s, after studying Bateman's paper in detail, she decided it was time to do exactly that. What she and her colleagues at the University of Georgia, Rebecca Steinichen and Wyatt Anderson, found contradicted Bateman in the most fundamental way. "We observed the movements of females and males in vials during the first five minutes of exposure to one another. Video records revealed females went toward males as frequently as males toward females; we inferred that females were as interested in males as males in females," they wrote in their paper, published in the journal *Evolution* in 2002.

This raised the dilemma of just how Bateman managed to see what he claimed to see in his own fruit flies. Investigating further, Gowaty soon began to notice problems with Bateman's study. In a subsequent paper, published in 2012 in *Proceedings of the National Academy of Sciences*, Gowaty and researchers Yong-Kyu Kim and Wyatt Anderson at the University of Georgia, wrote, "Bateman's method overestimated subjects with zero mates, underestimated subjects with one or more mates, and produced systematically biased estimates of offspring number by sex." They claim that Bateman counted mothers as parents less often than fathers, which is a biological impossibility, since it takes two to make a baby.

Another error is that the same genetic mutations Bateman needed his flies to have so he could distinguish the parents from their offspring also affected the fruit flies' survival rates. A fly with two severe and debilitating mutations, such as uncomfortably small eyes and deformed wings, could have died before Bateman had the chance to count it. This would have almost certainly skewed his results, too.

The mistakes are so clear, claims Gowaty, that Bateman's 1948 paper could only have been published if the editor—who should have checked for errors—hadn't actually read it. Failure to replicate scientific findings is a big deal. Often it leaves grave doubts about the original experiment. And for an experiment as important as Bateman's it should cause enormous concern.

In this case, though, the reaction to her findings has been mixed. "A lot of people were very excited about it, other people were pissed about it. . . . It was like they were mad," she tells me. When I e-mail Don Symons, who wrote *The Evolution of Human Sexuality* in 1979, to ask his opinion on Gowaty's failure to replicate Bateman's findings, he tells me he hasn't read her paper. When I ask instead for his broader thoughts on the evidence of multiple mating in females, he tells me that he's no longer available to answer my questions for personal reasons.

I also ask Robert Trivers, who first popularized Bateman's paper in 1972, for his response. "I was afraid you were going to ask that," he tells me over the phone from Jamaica. "I have not read the God Jesus paper." He agrees to look at it for me, but doesn't get around to reading it thoroughly even after a few weeks. "Since Patty is a careful scientist my bias is that she is correct," he finally tells me by e-mail. Even so, he adds that research on other species (including his research on a Jamaican giant lizard) has reinforced Bateman's principles. He sends me a paper published a couple of months earlier in the journal *Science Advances* by a team of European and US researchers.

It reviews examples from more than a century of animal data, concluding, "Sexual selection research over the last 150 years has not been carried out under false premises but instead is valid and provides a powerful explanation for differences between males and females."

For Gowaty, this defense isn't enough. Picking out examples in the animal kingdom that happen to be consistent with Bateman's principles ignores the wealth of inconsistencies—including, it seems, fruit flies. If there is enough contradictory evidence, this should put the underlying theory in doubt. The principles can't be considered principles if there are so many exceptions. The problem is that Bateman's and Trivers's ideas have taken on such a life of their own that this no longer appears to make much difference. "I think people are hung on Bateman's principles. They say that the principles stand whether the data are right or not," says Gowaty.

The failure of prominent scientists such as Symons and Trivers to read her work when it was published makes it even more difficult for Gowaty to make the wider scientific community aware of her findings.

"I find it tremendously strange," says animal behaviorist Zuleyma Tang-Martínez. "When a paper like that comes out, you would think that people who are interested in the topic would read it, regardless of which side they're interested in or which side they tend to agree with. I try to read papers by people who don't agree with my position. And I can't imagine just saying, 'Oh, I didn't bother to read it.' That to me seems almost insulting to a fellow scientist, to take that attitude."

For Gowaty, this is more than a professional frustration. "I think that our inability to see alternatives is associated with our commitment to see sex differences. The canon of sex difference research is about sex roles and the origin of sex roles and the fitness differences that supposedly fuel those. These arguments are the ones that we really need to understand in order to make inferences that are reliable. I happen to think that the canon is flawed, and it's flawed because it starts with sex differences to predict other sex differences. It is essentialist," she explains.

"Many of these theories that we have in evolutionary biology about sex differences are not fundamental theories. They're hand-wavy as hell."

That's not to say Bateman was completely wrong. Only that he wasn't entirely right. If we were to judge Angus John Bateman's principles today, it's likely that the jury would be out. "I think, certainly, there are species that fit that mold," says Tang-Martínez. In a review of evidence she published in the *Journal of Sex Research* in 2016, she lists the red-backed spider, pipefish, and seed beetles as examples of creatures that support Bateman's hypothesis.

"But I do think that given the amount of evidence all the way across the board, from male investment and cost of sperm and semen, all of the sort of original underpinnings of his whole idea, that we have to rethink," Tang-Martínez adds. "We can't just go on pretending that everything is hunky-dory, and that we can still apply Bateman across the board to all species."

She describes his principles as a box. As time wears on, fewer species—including humans—seem to fit in the box. Indeed, it's possible to argue that if ever there was proof that females aren't naturally chaste or coy, it's the extraordinary lengths to which some males go to keep them faithful.

———

"Let me tell you one anecdote from birds," Robert Trivers tells me.

It's from his graduate student days, when he would watch the pigeons on the gutter outside his third-floor window. In the winter, the birds would huddle together in rows for warmth. "You have two couples sitting next to each other in winter. They may have sex in December, but it's nonrepro-ductive, trust me. Throughout the winter they're not having sex together; they're just staying together, and they intend to breed together in the spring as soon as breeding season arrives," he begins.

The issue for the males is how to make sure they don't lose their female partners to another male. Trivers imagines himself as one of the male pi-geons. "If you have four individuals sitting next to each other, then the males sit on the inside, even though they are the more aggressive sex," he explains. "I sit in between the other male, who sits to my right, and my female sits to my left. He, meanwhile, has his female to his right. So both of us can relax during the night. We're in between any other male and our female." This arrangement means that each male can successfully protect his female from unwanted attention from the other male in their huddle.

But a dilemma sets in when another couple is added to the mix. With three males and three females, things get complicated. "Now it's impossible to have a seating arrangement such that each male is between his female and all the other males," he says. "So what you get instead is the outer two, the far left male and the far right male, each have their mate on the outside of them. So they're protecting their mate from contact with the other males." This leaves one male in a quandary. "Now, what about the central male? What does he do?" he asks me. "What he does is he pecks his female and forces her to sleep on the slanting roof several inches above him and several inches above the seat she would prefer to be on, which is sitting on the gutter, on which she would have a male on both sides of her." The male forces her to sit alone uncomfortably in the cold.

As a student, Trivers would sometimes work until three in the morning. "So at one thirty, I would hear some 'woo hoo-hoo,' and I would see, ha ha! What happened is the male has fallen asleep and the female has crept back down to the comfortable position, which is how she would prefer to sleep the night. He wakes up and sees she is there, and pecks her back up into this uncomfortable position!" he says. "The sexual insecurity or the risk of an extra-pair copulation is strong enough to make me willing to inflict a cost on my mate."

This phenomenon may seem bizarre—cruel, when seen through human eyes—but it's common across many species, including our own. It's known as "mate guarding." It's a vitally important piece of the puzzle when it comes to understanding relationships and the balance of power between females and males. Even though it might well harm the male to have his partner so distressed through the winter, leaving her with less energy come spring when she would need to reproduce and look after their offspring, he doesn't stop pushing her away from the other males. It's more important to him that he doesn't lose her to another pigeon, even for a moment.

For Trivers, this is powerful evidence of intense male competition for females. But seen from a different point of view, it also casts the underlying assumptions of Charles Darwin and Angus Bateman in an alternative light. Male sexual jealousy, the fear of being cuckolded, and such vicious mate guarding suggest that females aren't naturally chaste or passive at all. If they were, then why would their partners go to such extraordinary lengths to stop them getting anywhere near other males?

CHAPTER 7

Why Men Dominate

It cannot be demonstrated that woman is essentially inferior to man because she has always been subjugated.

—Mary Wollstonecraft,
A Vindication of the Rights of Woman, 1792

"I asked my mum to be cut," says Hibo Wardere, a forty-six-year-old woman from Mogadishu, Somalia, who now lives in east London. She was age six at the time, she continues, as we sit in a small, dark café near her home. She had no idea what she was asking for, of course, only that the other girls were bullying her for being the last one left. They told her she was dirty, that she stank. So she begged her mother for a procedure that, little could she have known as a small child, would cause her unimaginable pain and lifelong trauma: female genital mutilation.

Cutting of young girls is the norm in Somalia. There's a belief, says Wardere, that the practice dates back to ancient Egypt, when male slaves were routinely castrated before they worked in the households of the pharaohs. Nowadays it's common through large swaths of Africa and a few corners of the Middle East. The countries with the worst records include Egypt, Sudan, Mali, and Ethiopia, along with Somalia, where barely a girl escapes the knife. The United Nations World Health Organization estimates that more than 125 million women and girls alive today have undergone female genital mutilation in the countries where it's most concentrated, and almost all became victims before the age of fifteen.

The mutilation itself can take many horrifying forms. But the most common cuts fall into three categories. The first is the partial or total removal of the clitoris. The second includes this, plus the partial or total removal of the smaller, inner folds on either side of the vaginal opening. The third is the wholesale narrowing of the vagina's entrance by cutting and sealing the folds on either side, like a pair of lips being hacked and sewn shut. This final type, known as "infibulation," is often the most damaging of the three, leaving women with only a tiny gap through which to pee and pass menstrual fluid. It can be so small that they sometimes have to be cut open before they can have sex or give birth.

Infibulation is what was done to Wardere.

It happened forty years ago, but she remembers it as vividly as if it had been this morning. She grew up assuming that being cut was something to be proud of. It was a feeling reinforced when her female relatives threw a party in her honor to celebrate the big moment. They cooked her favorite food. They told her she was about to become a woman. In her six-year-old innocence she excitedly imagined that this might mean finally trying on her mother's makeup. "They made you feel like something amazing was going to happen," she tells me. "It was not like that. It was the beginning of a nightmare."

In Somalia, female genital mutilation is often carried out by a respected female elder, who's likely to have cut hundreds of girls already. Wardere recalls the woman who did it to her. "Her eyes haunt me even today. She instructed my mother, my aunties, and other helpers to hold me down, and they did. My mother looked away, but the others did hold me down. Then she ripped my flesh as I screamed and struggled and prayed to die. She just kept on going. It didn't bother her that I was just a child. It didn't bother her that I was begging for mercy." Wardere's torn flesh lay on the floor. The life sentence had been served. The cut was cruel enough, but she would also suffer recurrent urinary infections and scarring. The flashbacks would haunt her forever.

An entire decade would pass before she finally understood the point of it all. She never stopped asking her mother why she had allowed her to be cut. When she was sixteen, she was told that it was to put her off having sex before marriage.

For many millions of women, the agony of infibulation is quietly absorbed as an unavoidable part of life. In this silence, the practice continues to be inflicted on the next generation, the one after that, and so on, as it has for millennia. But Wardere refused to accept what had been done to her. "I decided I can't keep quiet," she says. When she arrived in England in the late 1980s, age eighteen and alone, fleeing civil war in Somalia, one of her first decisions was to seek medical help so she could be opened up.

She went on to marry happily and have seven children. In the last few years she's taken the brave step of speaking out about her experiences, and even detail them in an autobiography, *Cut: One Woman's Fight Against FGM in Britain Today*. As a prominent activist, she talks regularly in schools about the risks of genital mutilation and to urge girls not to become victims like her. This hasn't come without a price: Wardere has lost friends. When it was revealed that she refused to have her daughters cut, people warned her they would be considered impure. "They said nobody will marry them, that they're sluts."

The puzzling thing about female genital mutilation is that there seem to be no winners. Not men, not women. Wives have reported depression and domestic abuse because their husbands can't accept that they don't want to have sex. One young man admitted to her that he couldn't bring himself to sleep with his wife on their wedding night because she had undergone infibulation and he was scared of hurting her. If men would accept brides who weren't mutilated, she notes, the stigma might go away. Yet, however damaging it might be to their wives and their marriages, few men stand up against the practice.

And the reason for this is simple. The torture continues because it does what it was always intended to do. A woman who has been cut as a child will almost certainly remain a virgin when she's older. It would be too painful for her to be anything else. And once she's married, a husband can be confident that she'll be a reliably faithful wife. Throughout history, mutilating a girl's genitals has been the most viciously effective means of assuring a man that his children will be his own and not someone else's. It's as brutal a manifestation of sexual jealousy and mate guarding as anyone has ever seen.

The practice has been absorbed into some cultures so fully and for so long that women now have little choice but to give it their full cooperation. Without it, they risk being ostracized. Girls put pressure on each other to be cut, like they did when Wardere was six years old. Mothers take their own daughters to be cut, like Wardere's did. And female elders do the cutting. "It's all instigated by women. Men have nothing to do with it. But who are they doing it for? That's the question," she tells me. "It's all about control. They don't trust you with your own body."

In the café where we're meeting, older Somali men sit at neighboring tables sipping their coffees. She speaks loudly, refusing to be cowed. "They are doing it for him! It's all about him, it's not about you."

"A decent girl won't roam around at nine o'clock at night."

Female genital mutilation is only one way in which a woman's sexual agency is repressed. There have been countless others throughout history. The agonizing practice of foot binding, which is thought to have begun as a fashion fad in Imperial China in the tenth century, persisted into the twentieth. Young girls' feet would be so tightly wrapped in cloth that their toes would curve inward, leaving a pointed stump as tiny as three inches long. Historian Amanda Foreman has described how foot binding became a symbol of chastity and devotion in a society that prized obedience to men, centered on the teachings of the philosopher Confucius. "Every Confucian primer on moral female behavior included examples of women who were prepared to die or suffer mutilation to prove their commitment," she writes in *Smithsonian Magazine*. Like infibulation, it became so integral to Chinese culture that women became the mistresses of their own oppression. It was finally eliminated under pressure from China's Communist Party in the 1950s. There are a small number of older women alive even today with deformities caused by it.

As old forms of torture disappear, new ones swiftly roll in. In Cameroon and some parts of West Africa, girls between the ages of eight and twelve today suffer a procedure, often at the hands of their mothers, known as breast "ironing." A grinding stone, broom, belt, or another object is heated, then used to press a girl's budding breasts flat. The goal is to keep her looking like a child for as long as possible, so people assume she hasn't yet entered puberty. Aside from the psychological impact and immediate pain, breast ironing can cause long-term medical problems including scarring and difficulty breast-feeding, according to Rebecca Tapscott, who documented the practice for the Feinstein International Center at Tufts University in 2012.

Some methods of control, meanwhile, are deceptively subtle. Women in traditional Dogon communities in Mali use "menstrual huts" to seclude themselves during their periods. Beverly Strassmann at the University of Michigan in Ann Arbor and her colleagues discovered through field research, including many hundreds of paternity tests, that men who followed the traditional Dogon religion were four times less likely to be cuckolded than Christian men, whose wives didn't use the huts. It suggests that menstrual huts have allowed men to covertly track their wives' fertility.

Primatologist and anthropologist Sarah Blaffer Hrdy believes that all this—the systematic and deliberate repression of female sexuality for millennia—is what really lies behind the myth of the coy, passive female. She raised this, somewhat controversially, in her 1981 book *The Woman That Never Evolved*. Stepping outside the usual bounds of biology and viewing human behavior from a historical point of view, she asked whether scientists had approached the question of women's sexuality entirely the wrong way. Could it be that women and their evolutionary ancestors weren't naturally passive and monogamous, with a tiny sex drive, the way Charles Darwin and Angus John Bateman had assumed? Might it instead be the case that for thousands of years women had been compelled by men to behave more modestly?

Sexual jealousy and mate guarding are powerful biological drives seen throughout the animal kingdom, as biologist Robert Trivers learned in his observations of pigeons from his Harvard University window. If behavior like this had been exaggerated by humans, woven into society and culture, it might explain why women now appear to behave as modestly as they do. Like the female pigeon uncomfortably pecked back into her place by her mate, women may not be naturally passive and coy at all but just constrained in the ultimate interests of their mates. According to Sarah Hrdy, this explains the mismatch between science's old assumptions about female sexuality and the broad range of sexual behavior we actually see.

Her point is reinforced by the ways in which women are treated around the world. Besides horrific practices like female genital mutilation, few places exist that don't exercise a moral double standard. Passersby tut at the teenager who dares to bare too much flesh. Neighbors whisper about the single mother whose children have different fathers. From how she dresses and carries herself to how promiscuous she is, most societies expect a woman to behave more modestly than a man.

When this standard isn't enough to limit her behavior, humans have gone to elaborate lengths to enforce it. The most aggressive include forced marriage, domestic violence, and rape. One member of the gang who violently raped and killed a student on a bus in India in 2012 claimed to the BBC in an interview from prison that it was her own fault for taking the bus in the first place. As far as he was concerned, she was the one who had transgressed. "A decent girl won't roam around at nine o'clock at night," he told the reporters. "Housework and housekeeping is for girls, not roaming in discos and bars at night doing wrong things, wearing wrong clothes."

This double standard is even written into the laws of some countries. In Saudi Arabia, women's sexual freedom has been effectively removed because of the long list of things they're forbidden to do, including driving, mixing with men in public, and traveling without a chaperone or a man's permission. Although this takes repression to an extreme, the expectation of female modesty runs through many major religions. The hijab and burka worn by some Muslim women are demonstrations of this. The orthodox Jewish concept of *tzniut* similarly requires both sexes to cover up their bodies, but for married women in particular to cover their hair.

For Sarah Hrdy, the way female modesty is so deeply entwined with human culture like this, even to this day, has its roots in the ancient sexual repression of women. When developing this idea, she originally took her cue from a feminist psychiatrist called Mary Jane Sherfey, who had studied in the 1940s under Alfred Kinsey, the sexologist famous for overturning popular assumptions about people's sexual behavior. In 1973 Sherfey published an incendiary work of her own, exploring female orgasms. It was entitled *The Nature and Evolution of Female Sexuality*. Her conclusion was that the female sex drive had been vastly underestimated, and that women are in fact naturally endowed with an insatiable sex drive. Sherfey added that society itself was built around the demand to keep women's sexuality in check.

She wrote, "It is conceivable that the *forceful* suppression of women's inordinate sexual demands was a prerequisite to the dawn of every modern civilization and almost every living culture. Primitive woman's sexual drive was too strong." Its enormous strength was matched only by the incredible force that men through history had deployed to restrain it.

Unfortunately for Sherfey, she was largely dismissed by the scientific establishment, partly because her bold deductions went a little too far against the grain, but also because she made genuine scientific and anatomical errors. Don Symons, the anthropologist who has argued that the female orgasm didn't evolve for a purpose and that females have no biological reason to want more than one mate, was especially unimpressed. Sherfey's "sexually insatiable woman is to be found primarily, if not exclusively, in the ideology of feminism, the hopes of boys, and the fears of men," he wrote.

Sarah Hrdy, meanwhile, believed Symons was being unfair and that Sherfey, while wrong on many counts, had hit upon something important. Females *could* be sexually assertive. "Understand, Sherfey was writing years before primatologists knew much about sexual behavior in wild primates, certainly before we guessed at the existence of orgasmic capacity in nonhuman females; yet Sherfey's wild hunches anticipated future discoveries," Hrdy wrote in *Human Nature* in 1997.

The females of some monkey and ape species, we now know from a number of different sources, do appear to experience orgasms. In 1998 Italian researchers Alfonso Troisi and Monica Carosi published a paper in the journal *Animal Behaviour* describing orgasms in female Japanese macaques. They spent more than two hundred hours observing the monkeys in captivity, in which time they recorded almost the same number of copulations. In a third of these, females showed what they described as a "clutching reaction," which they interpreted as orgasm. This was associated with "muscular body spasms and, sometimes, characteristic vocalizations. When displaying the clutching reaction, the female arched her neck and/or reached back to the leg, shoulder, or face of the male and clutched his hair," Troisi and Carosi explained.

In the summer of 2016, evolutionary biologists Mihaela Pavlicev, at the University of Cincinnati College of Medicine, and Günter Wagner at Yale University, concluded that animal studies do indeed suggest that the female orgasm originated for a purpose. In their paper published in the *Journal of Experimental Zoology*, they outline how orgasms trigger a surge in hormones, which may in the past have been linked to ovulation—the release of eggs— as well as helping eggs implant in the uterus. Female cats and rabbits, for instance, actually need physical stimulation to release their eggs. In humans today, orgasms and ovulation aren't connected, but according to Pavlicev and Wagner, they may once have been.

By this logic, if orgasms aren't a vestige of male physiology and women really can have strong sex drives, then there must be another explanation for women being perceived as innately chaste and modest. Mary Jane Sherfey believed that something was holding women back from being the powerful sexual creatures they were born to be. This something was human culture.

———

Sherfey's line of thinking wasn't new. It stretched far back in feminist and political ideology.

"Couched in superstitious, religious and rationalized terms, behind the subjugation of women's sexuality lay the inexorable economics of cultural evolution which finally forced men to impose it and women to endure it," she wrote in *The Nature and Evolution of Female Sexuality*. "Generally, men have never accepted strict monogamy except in principle. Women have been forced to accept it." From the smallest laws to the most sweeping religious doctrines, she argued, cultures everywhere had tried to burn away every last scrap of female sexual freedom. This subjugation was the root of the moral double standard, the punishments, and the violent brutality that women continue to live with today.

In the nineteenth century, the German philosopher, journalist, and socialist Friedrich Engels, who famously collaborated with Karl Marx, had already drawn connections between the economic and political dominance of men and their control of female sexuality. He described it dramatically as "the world historical defeat of the female sex." He went on, "The man took command in the home also; the woman was degraded and reduced to servitude, she became the slave of his lust and a mere instrument for the production of children."

Just when in human history societies might have shifted from being fairly egalitarian to no longer equal is hard to pin down. Melvin Konner, anthropology professor at Emory University in Atlanta, tells me that when hunter-gatherers began to settle down and abandon their nomadic ways of living, between ten and twelve thousand years ago, things would have changed for women. With the domestication of animals and agriculture, as well as denser societies, specialized groups emerged. "For the first time you had a critical mass of men who could exclude women," explains Konner.

Systems of male control—patriarchies—emerged that exist to this day. And as they accumulated land, property, and wealth, it would have become even more important for men to be sure their wives were unswervingly faithful. A man who couldn't guarantee his babies were his own wasn't just being cuckolded but also risked losing what he owned. Mate guarding intensified.

Historian and feminist Gerda Lerner explored the subject in her landmark 1986 book *The Creation of Patriarchy*. Studying women in ancient Mesopotamia, a region spanning parts of what is modern-day Iraq and Syria, one of the cradles of human civilization, she pointed out that there was a strong emphasis on virginity before marriage. After marriage, a wife's sexual behavior was heavily policed. "Male dominance in sexual relations is most clearly expressed in the institutionalisation of the double standard in Mesopotamian law. . . . Men were free to commit adultery with harlots and slave women." Wives, by contrast, were expected to be completely faithful to their husbands.

Women, in no small way, were treated as the property of men. "Women's sexual subordination was institutionalised in the earliest law codes and enforced by the full power of the state," concluded Lerner. This included wearing the veil. Married, respectable women in the Assyrian empire in northern Mesopotamia, which existed until around 600 BC, were expected to cover their heads in public. Slave girls and prostitutes, on the other hand, were forbidden from wearing veils. If they broke this rule, they faced physical punishment.

Lerner suggested this subordination of women may even have given ancient civilizations their first model for slavery. "In Mesopotamian society, as elsewhere, patriarchal dominance in the family took a variety of forms," she wrote in *The Creation of Patriarchy*. "The father had the power of life and death over his children. . . . He could give his daughters in marriage . . . or he could consecrate them to a life of virginity. . . . A man could pledge his wife, his concubines and their children as pawns for his debt; if he failed to pay back the debt, these pledges would be turned into debt slaves."

Anthropologist Sarah Hrdy tells me, "Sexual jealousy is everywhere, even in nonpatriarchal societies. But it's so exaggerated in patriarchal societies because they're defending all these other interests." She has firsthand experience of how this feels. When she wanted to marry, Hrdy was forced to elope because some members of her conservative Texan family disapproved of her choice of husband. "Men still thought they had the right to tell me who I should marry. They thought they had the right to control my inheritance. They assumed that they owned me. Really it was about property, with women included as property."

Over thousands of years, this has had profound consequences for how women behaved and how they were then perceived. As patriarchies grew and spread, women gradually lost the power to earn a living, own property, lead a public life, or have much control over what happened to their children. The only freedom they were afforded were within the cages that had been created for them. So they were left with little choice but to behave in ways that served the system. A modest, coy woman who appeared to be chaste would marry well and prosper, while the less modest woman would be shunned.

As Sarah Hrdy shows in her own writings on the subject, there's plenty of evidence for this. Throughout recorded history, virginity and faithfulness have been universally celebrated as female virtues, and rigorously policed. In her 1999 book *Mother Nature*, she spans the globe with her examples. In India, there was the centuries-old Hindu practice of sati, in which widows sacrificed themselves (through choice, or not) on the funeral pyres of their husbands. Among the indigenous Maya people in southern Mexico and Central America there were terrifying tales of a demon who seizes and rapes women who behave immodestly. And in ancient Greece, women were taught to behave self-consciously through their dress and the way they carried themselves, their eyes downcast in the presence of men. "For the ancient Greeks, a woman's animal nature lurked at the core of her being. It was deemed necessary to 'tame' her," Hrdy writes. Aristocratic women, whose families had the most to lose by way of property and wealth, had practically no freedom at all. They were kept indoors, veiled, and in the shadows.

The shadow cast over women has never gone away. From the Mesopotamians to the ancient Greeks all the way to the present day, societies have restricted and punished women who have dared to breach the moral standard. By Charles Darwin's time, thousands of years into this regime, ideas of female nature had thoroughly adjusted to the new normal. Humans began to see women through a lens of their own creation. The job was done. Victorians, including Darwin, believed that women really were naturally coy, modest, and passive.

Female sexuality had been suppressed for so long that scientists didn't even question whether this modesty and meekness might not be biological at all.

"One of the first things that I noticed was that females were attacking males."

Even if humans once lived egalitarian lives long ago, was male domination of women inevitable? That's the question our complicated history and biology leave us asking. Does the biological drive that men have to guard females, combined with the fact that they're on average bigger and have greater upper body strength, mean that human societies would have always ended up with men in charge? Is patriarchy hardwired into our biology?

This may be impossible to answer, but science does have a perspective on it. The clues, some researchers believe, may lie in our primate past.

"The evolutionary perspective . . . reminds us that patriarchy is a human manifestation of a sexual dynamic that is played out over and over again, in many different ways, in other animals," writes anthropologist Barbara Smuts at the University of Michigan in a 1995 paper in the journal *Human Nature*. Smuts was known for her detailed field studies of monkeys and apes. She was a female pioneer in primatology, with many of her students themselves going on to have important careers in the field. This paper is particularly special, though, because it explores one of the thorniest aspects of our past: the possible evolutionary origins of patriarchy.

In her paper, Smuts details how far male monkeys and apes often go to sexually restrict the females of their species by force. Across the primate world, she explains, you can see evidence of male domination. When females are in the fertile phases of their sexual cycles, males tend to be far more aggressive. One example is the rhesus macaque, which lives in large troops. Males are about 20 percent bigger than the females. Researchers have observed that when a female macaque tries to mate with low-ranking males in the hierarchy, higher-ranking males try to block her by chasing or attacking her. Sarah Hrdy's observations of infanticide among Hanuman langurs in India are another example of males using violence to coerce females into mating with them. Mountain gorillas, according to Smuts, use the same tactic.

Hamadryas baboons in northern Africa are even more aggressive and "try to maintain control over females all the time," she writes. "When a female strays too far from her male, he threatens her by staring and raising his brows. If she does not respond instantly by moving toward him, he attacks her with a neckbite. The neckbite is usually symbolic—the male does not actually sink his teeth into her skin—but the threat of injury is clear." Orangutans provide another striking example of male coercion. For them, resisted mating appears to be the rule rather than a rare exception. Half of matings take place after long and brutal struggles with females.

But one of the most interesting cases for those who want to better understand humans is the chimpanzee. Along with bonobos, chimps are our closest genetic relatives in the primate world. Different estimates have dated our last common ancestor to be living eight to thirteen million years ago (the last ancestor shared by humans and dogs, by contrast, was possibly as far back as a hundred million years), which means we have a great amount in common. Researchers have noted that chimps are hierarchical, and males can be ruthlessly vicious toward other males when they're trying to establish themselves at the top of the order. Males show aggression toward females, too, although this aggression is about sexual coercion and mate guarding.

According to research published in 2007 by a team of prominent anthropologists, including Martin Muller, then at Boston University, more aggressive male chimps manage to mate more than the less aggressive ones. Even a low-ranking male will become aggressive when a female refuses him. Barbara Smuts has noted that the primatologist Jane Goodall once saw a male attack a female six times in five hours in the desperate effort to get her to mate with him. "Chimpanzees have been characterized in terms of their intercommunity warfare, meat eating, infanticide, cannibalism, male status-striving, and dominance over females," according to Craig Stanford, a professor of anthropology at the University of Southern California, in a 1998 paper in *Current Anthropology*. He adds that female chimps can be described as "essentially reproductive commodities over which males compete."

If a scientist had only ever studied chimpanzees, he or she might conclude that this is the natural order of life for the great apes. It's alluringly easy to draw parallels between patriarchal humans and macho male chimpanzees.

But according to Barbara Smuts, scientists have to be careful about this. In her 1995 paper on the evolutionary origins of patriarchy, she points out that, despite all the male aggression we see in the primate world, females aren't helpless victims. They rarely submit willingly to male control. They actually have their own clever ways of exerting power over males. "Although male primates typically are larger than females, this does not mean that they always win when they have conflicts of interest with females," she writes.

And there's one particularly strong example of this. It's the other primate with which we share as close a relationship as we do with chimpanzees.

———

The bonobo enclosure at the enormous San Diego Zoo in California attempts to approximate as closely as possible conditions in the wild where these creatures are from: the forests of the Democratic Republic of the Congo. There's a high waterfall, steep canyons with sunny and shady corners, and ropes that mimic hanging branches. The baby of the group, a fluffy, black two-year-old, leaps from one end to the other, following her mother. One of the older females sits comfortably, chewing on a long twig and occasionally peering at visitors through the glass barrier. To my eyes at least, the animals seem content.

All, that is, except one.

"I think he's traumatized," says Amy Parish, a primatologist who teaches at the University of Southern California. She has been studying bonobos in captivity for twenty-five years, starting her career at the University of Michigan in the 1980s as a student of Barbara Smuts. Parish tells me that the unhappy bonobo is a male named Makasi. We watch him a little longer. He's crouched alone to one side, with an arm resting on his knee. He softly licks his hand, which seems to be hurt. When he slopes off after a while, looking somewhat cowed, he keeps his injured hand protectively close to his head.

Bonobos are unusual in the ape world for being a species in which females dominate, with the oldest females appearing to be highest in the pecking order. Attacks by females on males are quite common.

"In bonobos it's very important for males to have their mothers with
them for life," explains Parish. "We have this pejorative idea that when
males are especially close to their mothers, that they're momma's boys and
that's a bad thing. But in this case, unlike chimps—where males separate
really clearly from their mothers at adolescence in order to join the male
dominance hierarchy—in bonobos, males maintain their relationship with
their mother for life. She intervenes in his fights, protects him from violence; he gets to mate with her friends; he gets access to otherwise exclusive
female feeding circles. So there's a really big upside for males."

Makasi's injuries were caused by a female called Lisa. "A good portion
of his finger is completely raw and the skin is gone. Several of his toes have
parts missing, and apparently he has other injuries. . . . But it's not uncommon that when males are injured that also there are injuries on the testicles
or penis or anus," she tells me. "Poor Makasi here was nursery reared. He
doesn't have a mother in the group who's willing to protect him, so he's vulnerable all the time. So he has very good reason to be cowed and afraid and
to keep his distance. To be careful."

Parish originally began studying bonobos to understand the role of
friendship between male and female primates. Barbara Smuts had done
similar work in baboons, but bonobos were something of a mystery. Until
1929 they weren't even understood to be a separate species from chimpanzees. Many decades later, when they were finally studied up close, bonobo
behavior turned out to be utterly different from that of their chimp cousins.
"For forty years, the chimpanzee researchers had the corner of the market
on man's closest living relative," explains Parish. "We built all our models of
evolution based on a chimp model: patriarchal, hunting, meat eating, male
bonding, male aggression toward females, infanticide, sexual coercion."
Bonobos turn this all on its head.

"One of the first things that I noticed was that females were attacking
males," Parish continues, as we sit on a bench next to the bonobo enclosure.
"Every zoo would have some explanation. Like, oh, this male bonobo was ill
when he was a youngster, and a female keeper took him home to nurse him
back to health. And she must have somehow ruined him, made him soft or
spoiled him. There was a zoo in Germany that didn't even believe human
females were suitable keepers for apes. Every zoo had some kind of folkloric

explanation for what was 'wrong' with their male, because it didn't seem like that was the proper way for males to behave, or for females for that matter. It seemed like a reversal of the natural order of things."

Parish decided to look through veterinary records at different zoos to see how widespread this phenomenon was. Serious injuries are always recorded, making it easy to spot any patterns. "It was just astoundingly in one direction," she tells me. In a group where there were multiple females, "females were systematically . . . inflicting routine, blood-drawing injuries on the males in the group." Evidence from the wild backs up the idea that bonobo females tend to hold the balance of power. As well as being dominant, they seem to mate freely with males from other groups, without fear of males in their own.

"I realized that these kinds of folkloric explanations in the zoos were probably not the real explanation," says Parish. "That maybe the 'natural' pattern in bonobos would be that females are dominant over males, and that instead of a patriarchy it was a matriarchy."

It was a radical suggestion. The word *matriarchy* has to be used advisedly. In bonobos, there are strong connections between unrelated females, and a matriarchy usually refers to networks of females who are related to each other. "When I proposed this idea in my paper, particularly the chimpanzee researchers were reluctant to accept that it might be true," she says. Some still resist the idea that females can be dominant in the same way males are in other species. Female bonobos have been labeled as "troublesome," Parish laughs, while males have been called "henpecked." Others have told her that bonobo males aren't dominated by females at all, but that they're somehow deferring to them in exchange for benefits like sex.

It's now widely accepted, though, that bonobo females do tend to dominate males. In this, they aren't alone in the animal kingdom. Female elephants are another more well-known example. They make up stable, core groups into and out of which males move transiently, depending on the breeding season. Spotted hyenas also live in clans ruled by an alpha female. Adult males rank lowest and eat last and are smaller and less aggressive than the females.

Aside from dominance, another way in which bonobos mark themselves apart from chimpanzees is in their sexual behavior. For the relatively brief time that I watch them at San Diego Zoo, I see three or four brief, casual copulations. This is quite normal. Bonobos seem to use sex as a kind of everyday social glue. Males have sex with males, females have sex with females.

Dutch primatologist Frans de Waal, based at Emory University in Atlanta, Georgia, who has worked closely with Parish, has described how
bonobos also engage in oral sex, tongue kissing, and genital massage. "Sexual
interactions occur more often among bonobos than among other primates,"
he writes in an article in *Scientific American* in 2006. "Despite the frequency
of sex, the bonobos rate of reproduction in the wild is about the same as
that of the chimpanzee. A female gives birth to a single infant at intervals
of between five and six years. So bonobos share at least one very important
characteristic with our own species, namely, a partial separation between sex
and reproduction."

One more difference is hunting. Female bonobos are usually the ones
who hunt for meat, often forest antelope, Parish tells me. "They flush the
young ones out amongst the tall grass, while the moms are off feeding, and
they eat them. There are some reports of males under the branches of females, throwing temper tantrums because they so desperately want some of
the meat and they can't have any of it unless one of the females, usually the
mother, wants to give them some. Or they can offer females sex in exchange
for food."

According to Parish, bonobo society works the way it does because females form powerful bonds with each other, even if they aren't related. "The
males can be friendly. They have sex with each other. But it's nothing like
the intensity or the scope that we see in the females. They sit together, play
chase and play wrestle, groom, share food, and have sex." The males are
usually physically larger, but by virtue of their tight bonds, bonobo females
manage to take charge. Observing the bonobos in San Diego Zoo, she found
that of the time females spent affiliating with other bonobos, two-thirds was
with females. De Waal has even described female bonobos as a "gift to the
feminist movement."

Their observations, though, still have a few critics. Chimpanzee expert
Craig Stanford argues that animals in captivity don't behave exactly the
same as those in the wild, because they're artificially forced into proximity
with each other. "I've never seen a wild bonobo, and I work on chimps, but
those of us who do fieldwork with great apes have tended to be a little skeptical of the view of those folks who say chimps are from Mars and bonobos
are from Venus," he tells me. "All of the female bondedness, female empowerment and sexuality, and all that stuff happen in a much higher rate
and in a much more prominent way in captivity then it does in the real
world, in the wild."

Parish disagrees. Although she has only studied bonobos in captivity, she insists, "There's nothing we see in captivity that hasn't also been documented in the wild. Sometimes the weights are different because they have more free time on their hands in a zoo. They don't have to go get their own food. But the repertoire is the same." Animal experts Sarah Hrdy and Patricia Gowaty tell me they agree that bonobos are today widely accepted as being an unusually female-dominant species.

_____

The stakes are high.

Primate research is high profile because of the enormous implications it could have for how we understand human evolution. It's tempting to want to categorize ourselves as being either like chimps or like bonobos, because the two species so neatly encapsulate the modern battle of the sexes. Judging humans by our patriarchal history, it's easy to see why so many primatologists have compared us to chimpanzees. But is it possible that somewhere in our evolutionary history we were matriarchal like bonobos appear to be?

For primatologist Amy Parish, the existence of a primate species in which females tend to dominate is hugely important, if only because it opens the debate. "When we only had chimps in the model, it seemed like patriarchy was cemented in our evolutionary heritage for the last five to six million years, because we share so many traits in common with them. The kind of 'man the hunter' model, all of that was based on chimps. Now that we have an equally close living relative with a different pattern, it opens up the possibilities for imagining that it's possible in our ancestry that females could bond in the absence of kinship, that matriarchies can exist."

Bonobos aren't the only primate species in which females cooperate. Hanuman langurs, as documented by Sarah Hrdy, for example, band together to fight off outside males intent on killing their infants. Some female primates are also known to use social relationships with males to defy control, according to Barbara Smuts. In one baboon group she studied in Kenya each female had a "friendship" with one or two males. "Friends traveled together, fed together, and slept together at night," she explains. The male friend would protect her and her infants from other males, which meant that she faced less harassment. It's an arrangement that prompted Harvard University primatologist Richard Wrangham to describe these males as "hired guns."

The focus on dominance in primate behavior makes it easy to forget that there are also species out there in which the sexes coexist and cooperate relatively peacefully. Pair-bonded tamarins and titi monkeys, for instance, share child care between males and females. Titi monkeys don't seem to have any kind of dominance hierarchy. In other monogamous species, such as gibbons and simiangs, male coercion of females is hardly seen.

A common mistake is to assume that males naturally dominate because they're larger. And this makes intuitive sense. If any one sex can take control, isn't it likely to be the one with the physical advantage? But this isn't true. Gibbons of both sexes look similar, for example, but the males tend to be very slightly larger and don't coerce females. Size is a product of many factors, including the need to physically outstrip opponents in the competition for mates. For females in particular, not all their energy can be driven into height or size because they also need it for reproduction and lactation. There's not always a correlation between size and male dominance over females.

Indeed, Katherine Ralls, a zoologist and researcher at the Smithsonian Institution in Washington, DC, confirmed this all the way back in 1976: "Females are larger than males in more species of mammals than is generally supposed." In her paper in the *Quarterly Review of Biology* she adds that, for a variety of species, size doesn't seem to correlate reliably with which sex is dominant. The African water chevrotain, which is a type of deer, and many small antelopes, for example, have larger females who aren't dominant. Meanwhile, the Chinese hamster, ring-tailed lemur, and pygmy marmoset all have smaller females that dominate the males. Bonobo females, too, are generally smaller than the males. "Their larger size is balanced by the fact that females cooperate against males, whereas males seldom cooperate against females," notes Barbara Smuts.

The common thread that unites species in which females are particularly vulnerable to male violence is females being alone. An orangutan female, for instance, will travel alone with her dependent young almost all the time. Female chimpanzees, adds Barbara Smuts, spend three-quarters of their time alone, with no other adults present.

Human life is far more complex, of course. It can't be generalized the way life in other species often can. But in this respect at least, we appear to parallel each other. In patriarchal societies, a woman will almost always leave her own family when she gets married and go live with her husband's. Losing the support of her relatives makes her especially weak in the face of violence and repression. And this weakness is exacerbated when men form alliances with each other and control resources, such as food and property.

In the end, this is where the die seems to fall when it comes to male dominance over females. Female cooperation makes the difference. This doesn't answer the question of whether male domination was always the biological norm for our species, the way it is for chimpanzees, but it does offer a perspective on the battle for equality today. For Amy Parish, the great apes are not just a window on our possible past but also an example of the different ways we could live in the future. Her work shows that male domination isn't inevitable when females work together to establish their interests—the way that bonobos do.

"It's certainly given me hope for the human feminist movement," she tells me. "That here we can see females actually bonding with each other, maintaining those bonds, maintaining that loyalty. And then ultimately having the power in their groups. So I think they're a great model for that. That yes, females can be in charge. They can control the resources. They don't need to go through males to get them. They don't have to be subjected to sexual violence or infanticide, all because they have the upper hand. And they do that by staying loyal to their female friends."

CHAPTER 8

The Old Women
Who Wouldn't Die

*Women may be the one group that grows more radical
with age.*

—Gloria Steinem, *Outrageous Acts
and Everyday Rebellions,* 1983

I am at the end of my research, and I've reached Bedlam.

I'm only visiting—I want to better understand the experiences through
history of women going through menopause—but this place makes me un-
easy nonetheless. Bethlem Royal Hospital is one of the oldest psychiatric
institutions in the country. It has shifted sites around London three times
since it was established in 1247. Along the way it acquired such a shocking
reputation that its very name, shortened to Bedlam, became synonymous
with chaos and uproar. Things got so bad in the nineteenth century that the
government carried out inquiries into patient abuse, which forced reforms
of the hospital.

An article by one doctor in the *British Medical Journal* in 1912 states that
one in twelve of the women being admitted to insane asylums and hospitals
like these across the country at that time were postmenopausal. In private
institutions, where the wealthier tended to go, they were one in ten. The
hormonal and physical changes associated with menopause, as well as the
shift it marked in their life and status as mothers, had impacts on the mental health of many older women. Some cases were documented with medical fascination. One doctor described a forty-nine-year-old woman who
believes she's decaying. She eventually commits suicide. Another, age fifty,
complains that she's no longer a human being, with no stomach, heart, or

lungs. A forty-six-year-old wife, meanwhile, develops the habit of stripping naked and demanding sex.

This was a time when menopause was grossly misunderstood. Fairy tales painted women at the end of their childbearing years as useless, crazy old crones. They lived in shoes with too many children or killed innocents in gingerbread houses. Farther back in history, they had been treated as witches more literally. At the Salem witch trials in Massachusetts in 1692, sixteen accused women were executed or died as a result of their incarceration, and from what we know, at least thirteen of them were menopausal.

With little understood about menopause and the mental pressures facing older women, people in the nineteenth century tried out all sorts of disastrous cures for its symptoms. One was bloodletting, to get rid of what was believed to be unspent menstrual blood. Sometimes they were given drugs like opium or morphine. In the worst cases, women had their ovaries surgically removed. Meanwhile those who ended up in asylums like Bedlam may have found themselves in the care of strict, fatherly male doctors, bizarrely advising them to drink less alcohol, take hot baths, and wear flannel underwear. One doctor at the time even suggested that menopausal women retreat to a quieter life and withdraw from the outside world, reflecting the attitude that they should be neither seen nor heard.

Life in the asylum wasn't easy. A woman arriving at Bedlam between 1676 and 1815 would have been welcomed by two imposing stone statues flanking the entrance. They represented the two categories into which most mental patients were thought to fall. The first figure was *Raving*, desperately struggling against hospital chains, his face contorted with agony. The second, *Melancholy*, was unrestrained but disturbingly unengaged, as though the outside world had lost all meaning. Of the women admitted to Bethlem Hospital for menopause-associated mental illness, only up to half recovered, according to data in 1912.

Thankfully, the bad days of Bedlam are over. In its current incarnation on a picturesque country house estate in southeast London, Bethlem Hospital is a peaceful place. There's a collection of small, low wards, each in a separate building, nestled in hundreds of acres of soft greenery. *Raving* and *Melancholy* now live in the reception of a small, sunlit museum inside the grounds, where they are brought to life upstairs in the histories of real people. On the wall I find two nineteenth-century photographs, both of older women. One is suffering from chronic mania, her face faintly twisted as she grips a lifelike doll dressed in long white robes. The other, the caption says, suffers from melancholia. She looks as though she's reflecting on her life, with a pained, faraway look in her eyes.

If fertility represented youth and health, society assumed, then infertility was exactly the opposite. It wiped out the entire point of being female. It turned a woman into something else. And this was reflected in the ways older women were treated, especially by science and the medical profession.

"Estrogen-starved women."

In 1966 a sensational new health book was published in the United States, promising women that they had nothing to fear from growing old, because science could make them young again. It became an instant hit, selling a hundred thousand copies in just seven months. Its title was as seductive as its contents: *Feminine Forever.*

According to the author, New York gynecologist Robert Wilson, the answer to women's (and husbands') prayers came in the shape of sex hormones. With a youth-restoring blend of hormones including estrogen, he claimed, a woman's "breasts and genital organs will not shrivel. She will be much more pleasant to live with and will not become dull and unattractive." They couldn't reverse infertility, but hormones could at least swat away the hot flashes and mood swings that damaged some postmenopausal women's lives.

It sounded too good to be true. It wasn't—at least not entirely. Wilson wasn't a total quack. With the dawn of endocrinology in the early twentieth century, scientists had finally got a grip on what was actually happening during menopause. The biological mechanism turned out to be quite simple. Every month or so, ball-shaped pockets called follicles grow inside a woman's ovaries. They release the eggs that are needed to make babies and secrete estrogen and progesterone. Girls are usually born with somewhere between a million and two million follicles, although most of these are gone by the time they hit puberty. Over decades, all the follicles eventually disappear, and it's their loss that spells the start of menopause. This means no more baby-making eggs and also a drop in hormone levels.

The loss of estrogen in particular is what prompts the symptoms we usually associate with menopause, such as hot flashes, a change in sex drive, mood swings, and weight gain. Hormonal changes before the start of menopause usually begin around age forty-five, with menopause itself starting on average between fifty and fifty-two. It has been estimated that around 5 percent of women experience menopause early, before they're forty-five. By giving a menopausal woman extra hormones, as Robert Wilson advocated, some symptoms could be alleviated.

Indeed, hormone treatment had already been around for decades before his book was published. In the 1930s a small number of doctors and pharmaceutical companies had begun to reframe menopause as a disease of deficiency, like not having enough vitamins. In some parts of the world, it was no longer seen as a normal, natural part of aging. Within a few decades, it became almost routine for women to take estrogen pills or injections when they reached menopausal age.

According to Saffron Whitehead, emeritus professor of endocrinology at St George's, University of London, treatments boomed in the 1950s and 1960s. After the Second World War, women who had worked as part of the war effort in Europe were instead encouraged to be housewives, and the idea was that hormone therapy "would keep women sexy and at home," she explains. Ads for Estinyl hormone tablets from 1952, for instance, feature beautiful, smiling women, their faces floating serenely in a sea of flowers.

Robert Wilson chose to send his own message with a sledgehammer rather than with flowers. He argued that menopause should be recognized as a "serious, painful and often crippling disease," turning its sufferers into what he disdainfully described as "castrates." Anne Fausto-Sterling, professor of biology and gender studies at Brown University, who has written about his work, describes his disparaging depictions of "estrogen-starved women." They are portrayed as existing rather than living, she says. Pictures that he includes in one of his published papers show elderly women walking along in public dressed in black and hunched over. "They pass unnoticed and, in turn, notice little," he warns his readers.

By the 1960s the hormone treatment wagon had turned into a juggernaut. After *Feminine Forever* came out in the United States, British journalist Wendy Cooper saw similar success in the United Kingdom in 1975 with her book *No Change: Biological Revolution for Women*. "She said it was the best thing that had ever happened to her," recalls Saffron Whitehead. "Everyone, because of this publicity and how young it kept you, would take it."

———

Of course, no magic cure ever turns out to be as magic as it first appears. After Robert Wilson died, a scandal in 1981 revealed that his pockets had been lined all along by pharmaceutical companies who were trying to sell more hormone replacement drugs. His best-selling book *Feminine Forever* had been bankrolled by Wyeth Ayerst, one of the therapy's biggest manufacturers.

More worryingly for the many women who had been convinced of the transformative power of hormones, researchers discovered that there might be a dangerous link between estrogen replacement therapy and cancer of the lining of the womb. In the early 1990s large studies showed that hormone treatment mixing estrogen and progesterone increased breast cancer risks. And by 2002, another important study confirmed that estrogen and progesterone really weren't the panaceas they seemed. Hormone replacement therapy, while changing the lives of many women for the better, also increased the risk of heart attacks and strokes.

Prescriptions plummeted and women were advised to take the drugs only for serious menopausal symptoms. Hormone treatment remains a much-welcome boon to many women who take it, but doctors today tend to prescribe it for no longer than two to four years, says Saffron Whitehead. She herself took hormone therapy for fewer than three years. "We're now on the fence about it," she says, adding that scientists are still analyzing data to get a clearer grip on how safe it is.

The saga has a both a good and a bad ending. The medical drama surrounding hormone replacement therapy certainly caused uncertainty and panic, and risked lives. But it at least shone a much-needed light on the health of older women. Researchers have dedicated more time to picking apart what the symptoms of menopause really are and to better treating the other problems, including psychological illnesses, associated with old age. A few scientists are even working on solutions to help women of menopausal age get pregnant or delay the onset of infertility.

At the same time, other scientists have turned their attention to the bigger, broader evolutionary question of why women experience menopause at all. Does it serve a purpose, which has some biological logic to it? Or is it like wrinkling and gray hair—an unavoidable aspect of aging, a disease of deficiency—that mark the body's inevitable decline? Why then do all women experience it, but some men seem to be able to keep reproducing until they die?

When it happened to my mother, she remained an active, working woman. Menopause didn't stop her from running a business, teaching yoga, cooking, and babysitting. And her experiences are echoed by people like her the world over. As far as history tells us, it has always been this way. The existence of healthy postmenopausal women poses an enormous dilemma for evolutionary biologists. Why would nature render them infertile when they are still so full of life?

"These old ladies . . . were just dynamos."

When a phenomenon as important as menopause happens in humans, we almost always find it in other species, too, particularly among our primate relatives, like chimpanzees and the other great apes. But with menopause, that's not the case. It's freakishly unusual. In almost every species, females die before they become infertile. Chimpanzees are, like us, fertile for no more than around forty years. The difference is, in the wild they rarely survive beyond this. Elephants live longer but carry on having babies into their sixties. A long postmenopausal life is so rare that, as far as we know, humans share it with only a few distant species, including killer whales, which stop reproducing in their thirties or forties but can survive into their nineties.

The reason for this rarity seems to be that we and all other animals have few physical features that haven't been pared down by evolution to make them fit their purpose. We are streamlined by nature, having long ago ditched most of what we don't need and honed what we do. Life span appears to be one of those features. By and large, animals live long enough to have children and maybe see them grow up, and then they die. If you can't reproduce and your genes are no longer being passed along to another generation, then harsh as it seems, nature generally doesn't want to know. This logic dictates that nobody should be alive beyond menopausal age. By this ruthless measure, my mother and all postmenopausal women should be dead.

Yet they're all around us. What's more, on average, they live even longer than men do, even though men can keep producing sperm well into old age. (Although one study in 2014 did find that semen tends to change after the age of thirty-five, making it slightly less likely a man's partner will get pregnant after sex. Research published in 2003 also showed that pregnancies from older fathers, especially past the age of fifty-five, are more likely to lead to miscarriage and birth defects.)

Answers to the mystery began with a brief observation made back in 1957 by George Williams, a biologist who was one of the most important evolutionary scientists of the twentieth century and at that time working at Michigan State University. The exact question he was pondering was why women lose the ability to have babies in middle age so abruptly when other parts of aging happen more gradually. He proposed, briefly and without much elaboration, that menopause may have emerged to protect older women from the risks linked to childbirth, keeping them alive long enough to look after the children they already had.

Until fairly recently, having babies was a huge killer of women. In the nineteenth century the number who died from or during childbirth in England and Wales hovered between four and seven for every thousand births, and this didn't fall significantly until after the Second World War. Having babies into old age would have multiplied the risks for both mothers and their children. "It is improper to regard menopause as a part of the aging syndrome," concluded Williams. His kernel of an idea came to be known as the "grandmother hypothesis."

To parents whose own parents are still alive, the grandmother hypothesis makes instinctive sense. For me, sitting at my desk today is a benefit made possible by my mother-in-law. She's busy taking care of my baby son, leaving me free to do other work or have more babies. And she isn't alone. Grandmothers (and it has to be said, a few grandfathers, too, these days) are a common sight on the streets where I live in London, pushing buggies in the middle of the morning and carting beloved spawn of spawn back from schools and nurseries in the afternoon. It's a trend that we nowadays associate with busy working parents and the high cost of child care, but it has far longer roots. Extended families, in which children live with their grandparents, were until recently a common feature around the world. In Africa and Asia, they still are. Research by the US-based organization Child Trends found in 2013 that at least 40 percent of children in Asia live with extended family as well as their parents. This, in essence, could have been the kind of crucible in which the grandmother hypothesis operated.

The focus on grandmothering also casts menopause in a new light, suggesting that it isn't some biological blip or routine curse of old age, but that it's there for a distinct evolutionary purpose: to allow women to safely continue caring for their children as they grow older and perhaps also be there for their grandchildren. The old image of the useless crone is replaced by a useful woman. Rather than being a burden on society, retreating into a quieter life, she is front and center. She is propping up her family.

For the sixty years since Williams first shared this thought, researchers have been searching for the evidence to prove it.

———

"I was just trying to understand what the men were doing," says Kristen Hawkes, professor of anthropology at the University of Utah. She's the world's leading researcher on the grandmother hypothesis, and its strongest advocate.

Hawkes spent the 1980s doing fieldwork with the Aché, nomadic hunter-gatherers in Eastern Paraguay. And she soon realized, like anthropologists before her, that men weren't providing all the food for their families. Hunting by men alone simply didn't put enough on the table for women and children to survive. "The things that they were foraging for were the things that went around to everybody. So the fraction that went to their wife and kids was no different from what everybody else got," she tells me. Meat from a hunt not only had to be divided among many but was also sporadic. Long periods of time could go without a kill.

Trying to uncover more clues about how mothers and babies were surviving, Hawkes went to study Hadza hunter-gatherers in Tanzania. The Hadza are particularly special to anthropologists because they arguably live a life as close to how humans lived before agriculture as anyone is likely to find today. A large portion of the Hadza don't tend crops or herd animals, and they live south of the Serengeti in a region not far from where fossils have been found of some of our very earliest ancestors. "That was paramount for me going to the Hadza," she explains.

And it was there that she saw hardworking grandmothers.

"There they were, right in front of us. These old ladies who were just dynamos." It's impossible to hear Hawkes talk about her fieldwork and not get carried away by her excitement. Her voice shifts a gear and, to this day, she sounds genuinely surprised by what she found all those decades ago. There was a division of labor between childbearing women and grandmothers, with active older women foraging for food alongside everyone else.

Hawkes discovered that the Hadza grandmothers and other senior women, including aunts, helped daughters raise more and healthier children. They were vital to reproduction even if they weren't themselves having babies. Grandmothers, she suggested, were also the reason women were able to have shorter intervals between babies. They stepped in to help before other children became independent. Her landmark scientific paper on the subject, published with her colleagues in 1989, was titled "Hardworking Hadza Grandmothers." More work by Hawkes and her team has since revealed just how industrious they are. Women in their sixties and seventies are described as working long hours in all seasons, bringing back as much food or even more than younger women in their families.

Other anthropologists have seen similar things. Adrienne Zihlman, who helped develop the idea of woman the gatherer, recounts a particularly vivid example for me, which she read in the *New Yorker* in 1990. It comes from the writer Elizabeth Marshall Thomas, who lived with nomadic hunter-gatherers in the Kalahari in southern Africa. Thomas describes a group of people who fell ill during an epidemic. One young widow and her two children were too sick to leave with the group when it decided to shift camp in search of more food. "But her mother was there," she writes. "This small, rather elderly woman took her daughter on her back, her infant grandchild in a sling across her chest, and her four year old grandchild on her hip. She carried them thirty-five miles, to her people's new camp." The superhuman efforts of this grandmother meant her daughter and two grandchildren recovered from their illness and weren't left behind.

A common counterargument to the grandmother hypothesis, known as the "extended-longevity" or "life-span–artifact" hypothesis, is that menopause must be a by-product of our increased life expectancy. We don't have to go back many generations to know that we're living on average far longer and healthier lives than our ancestors. In the United Kingdom, life expectancy for women has risen from forty-nine years in 1901 to almost eighty-three in 2015. This is expected to go up by another four or so years by 2032. In the United States, female life expectancy was just over eighty-one years in 2015, according to the Centers for Disease Control and Prevention's National Center for Health Statistics. So the line of reasoning goes that older women become infertile because, were it not for all the good food they eat, sanitation, and modern medicine, they would be dead too early to experience menopause at all.

In reality, life-expectancy data can be misleading. A large chunk of a population's average life expectancy will often be decided by infant mortality. If more children die, this drives down the average. This in turn means there's every likelihood some people were long ago achieving ripe old ages, even if most of the people around them had shorter lives. According to the latest findings, it is almost certain that some women would have experienced menopause in our ancient past. The earliest recorded mention is often attributed to Aristotle in the fourth century BC, when he is supposed to have noted that women stopped giving birth around the ages of forty or fifty.

Research comparing the body weights and body sizes of our primate cousins suggests that a small proportion of our early human ancestors could have lived to between sixty-six and seventy-eight years. Most convincing, scientists studying hunter-gatherers the way Kristen Hawkes has done have noticed that between 20 and 40 percent of adult women are postmenopausal. In other words, older women would always have existed.

In her book *Mothers and Others*, anthropologist Sarah Blaffer Hrdy suggests, "Fewer than half of Pleistocene mothers would be likely to have had a mother alive or living in the same group when they first gave birth." So not every child would have had a grandmother alive, but many would have. Grandmothers are the "ideal allomothers," she adds. "Experienced in child care, sensitive to infant cues, adept at local subsistence tasks, undistracted by babies of their own or even the possibility of having them, and (like old men as well) repositories of useful knowledge, postmenopausal females are also unusually altruistic."

Hard data, too, have backed up Hawkes's findings. Studies in Gambia have found that the presence of a grandmother increases a child's chance of survival. Similar results have been found in historical data from Japan and Germany. One study of three thousand Finnish and Canadian women from the eighteenth and nineteenth centuries found that women had two extra grandchildren for every ten years they survived beyond menopause.

In 2011 evolutionary demographer Rebecca Sear and biomedical scientist David Coall pulled together research from across the world to find out who, other than mothers, has the greatest impact on child survival. They concluded in their paper, published in *Population and Development Review*, that maternal grandmothers are consistently among the most reliable helpers. "In more than two-thirds of cases her presence improved child survival rates. Paternal grandmothers were also often associated with positive survival outcomes, though somewhat less consistently: in just over half of cases they improved child survival," they note.

––––––––––

"Very few species have a prolonged period of their life span when they no longer reproduce," says Darren Croft, a psychologist who studies animal behavior at the University of Exeter in the United Kingdom. Croft has a particular interest in resident killer whales—orcas—one of a few species of whale in which females are known to stop having babies and yet live for many decades afterward, sometimes into their nineties. The males die far younger, in their thirties or forties.

His explanation for this, outlined in a paper published by him and his team in the journal *Science* in 2012, lies in the power of the enormously tight bond between whale mothers and their sons. "Female killer whales act as lifelong carers for their own offspring, particularly their adult sons," he explains. A mother killer whale with a son focuses her efforts on him throughout her life. Indeed, such is the connection between them that data have shown that when a mother killer whale dies, her son is more likely to die far younger. Incidentally, this is just a son thing. The link between the life spans of mothers and daughters is weaker.

Croft carried out further research with colleagues at Exeter University, York University, and the Center for Whale Research in the United States, published in *Current Biology* in 2015, also looking at resident killer whales in the northern Pacific. Watching the whales led them to believe that it's the wisdom gathered over their lifetimes that makes the older females so invaluable. "They are more likely than the males to lead a group of orcas, especially in times of short food supply," says Croft. "For killer whales, what's really crucial is when and where salmon is going to be," and the older females seem to have this knowledge.

Croft believes that research like his into menopausal whales, unusual though they are, could provide an extra piece of the human menopause puzzle. If this can happen in the wild to another species, then it could have happened to us. "Following old females isn't unique," he adds. Elephants, too, have matriarchs who seem to have special information about threats from predators.

Since the grandmother hypothesis has emerged, other theories have added to it. In 2007 Barry Kuhle, in the psychology department at the University of Scranton, proposed that fathers (more specifically, absent fathers) might also have helped in the evolution of menopause. His idea is that men become less involved parents as mothers get older, partly because they die sooner but also because they are more likely to leave their partners. This supports the grandmother hypothesis, again, because it makes what grandmothers do even more vital. "I simply added an additional factor," says Kuhle.

Others have added that grandmothers aren't necessarily heartwarming, selfless babysitters living in harmonious families. Research published in the journal *Ecology Letters* in 2012 has indicated that what forces some women into caring for their grandchildren is intergenerational conflict rather than the failure to have babies of their own. Evolutionary biologist Virpi Lummaa and her colleagues studied parish-record data in Finland and found that infant survival was drastically reduced when daughters-in-law and mothers-in-law had babies at the same time, if there weren't enough resources for all the children. If a mother-in-law cares for her grandchildren, she benefits because she is genetically related to them. There's no such benefit the other way round for the daughters-in-law, says Lummaa. Grandmothers are genetically related to their grandchildren, whereas daughters-in-law are not genetically related to their sisters- and brothers-in-law. Grandmothering, then, is just a canny choice when resources are scarce.

"Men, young and old, prefer younger women."

The grandmother hypothesis hasn't gone unchallenged.

At least a dozen competing ideas have come along over the years, each with its own drawbacks and merits. They include the follicular depletion hypothesis, which, like the extended longevity hypothesis, says that women nowadays outlive their eggs. The problem with this is that you might then expect women with more children to go through menopause later, because they're not menstruating while pregnant. They don't. Another hypothesis focuses on reproductive cost, saying that baby making takes such a large physical toll on a woman's body that menopause evolved to protect her from further damage. But if this were true, we might expect to see women with more children experiencing menopause earlier, and we don't. Another, the senescence hypothesis, offers up the possibility that menopause is just a natural feature of aging, like wrinkles or loss of hearing. And while other side effects of old age may happen gradually, including male infertility, female fertility just happens to end more abruptly for physical reasons.

In 2010 evolutionary biologist Friederike Kachel and a team of researchers at the Max Planck Institute for Evolutionary Anthropology in Leipzig, Germany, decided to run a test to see if the grandmother hypothesis really is the best explanation for menopause among these alternatives. They ran computer simulations of how humans could have evolved with women living longer after menopause. To the surprise of anthropologist Kristen Hawkes and her team, who by now had amassed many years of evidence in their favor, Kachel's group found that while helpful grandmothers certainly did raise the survival rates of their grandkids, this effect didn't appear to be enough to account for why women now live so long.

In 2012, rescuing the hypothesis from news reports that were already questioning it, Hawkes's team published the results of their own computer simulation, which showed that slowly increasing the proportion of particularly long-lived grandmothers in a population, from 1 percent to 43 percent over the course of thousands of years, could indeed drive up everyone's life span. She and her colleagues believe that part of the problem with the German mathematical model may have been that it was run for just ten thousand years, when in fact the long sweep of human evolution means that the effects could have taken much longer to appear. They also argue that the model hadn't accounted for possible costs for men in living longer, such as having to compete with the same number of fertile men for a proportionally smaller pool of fertile women.

Then, in 2014, Hawkes and colleagues at the University of Utah and at the University of Sydney, Australia, plugged their numbers into another mathematical model. In this one, they assumed that at some point in human history we all had similar life spans to our primate cousins, and that, like them, women died before menopause could kick in. The model then slowly feeds in a small number of women with genetic mutations that mean they live longer than everyone else. The mutation spreads, and eventually, very gradually, everyone is living longer.

"When you add helpful grandmothering, at the beginning, almost nobody is living past their fertility," explains Hawkes. "And yet just those few, those few who are still around at the end of their fertility, that's enough for selection to begin to shift the life history from an apelike one to a humanlike one. We end up with something that looks like just what we see in modern hunter-gatherers." All it would have taken in those early days of human evolutionary history was a few good grandmothers.

Not everyone accepts this.

———

"Let's assume mating is not random," evolutionary biologist Rama Singh tells me over the phone from McMaster University in Canada. It sounds as though he's smiling, aware of just how provocative his comments are going to be.

As both of us know, his is the most controversial countertheory to the grandmother hypothesis. "We know that men, young and old, prefer younger women. So in the presence of younger women, older women will not be mating as much," he explains. If they aren't having sex, his argument goes, they don't need to be able to reproduce. In summary, older women become infertile because men don't find them attractive. One reporter has described this account as putting the "men" in "menopause."

In 2013 Singh, along with two colleagues at McMaster, Richard Morton and Jonathon Stone, published the idea in the journal *PLOS Computational Biology*. It was the kind of paper that instantly attracted worldwide news coverage and a barrage of correspondence. "A lot of women wrote bad letters to us," admits Singh. "They thought we were giving men more say in evolution." One sarcastically demanded to know just how much sex she would need to have as an older woman to avoid menopause.

"Whether you believe it or not, just look around society today. The science is cut and dried," he responds, when I ask him about the criticism. "The truth is, nature doesn't care about sympathy or feeling."

Many, however, have challenged his view of nature. Indeed, Singh, Morton, and Stone's hypothesis has been mocked in scientific circles. "It makes very little sense. Chimps actually prefer older females as their mates," Virpi Lummaa, at the University of Sheffield, tells me. Another prominent researcher in the field, Rebecca Sear, at the London School of Hygiene and Tropical Medicine, agrees. "It's a stupid argument and it was trashed when it came out. It's a circular explanation. The reason men don't prefer postmenopausal women is that they're postmenopausal and they can't get pregnant, not the other way round." Even so, Singh and his colleagues have stuck to their guns, unapologetically.

Their idea isn't entirely new. The inspiration for it stems back to 2000 when anthropologist Frank Marlowe, then working at Harvard University, published a provocative explanation for menopause known as the "patriarch hypothesis." Like the name suggests, the patriarch hypothesis is about powerful men, specifically, men powerful enough to be able to have sex with younger, fertile women even as they get old. "Once males become capable of maintaining high status and reproductive access beyond their peak physical condition, selection favoured the extension of maximum life span in males," Marlowe explains in his paper, which was published in the journal *Human Nature*. Even a few high-status old men spreading their seed would have been enough to produce a difference in how long humans lived, he argues.

Since the genes associated with increased life span happen not to be on the Y chromosome, which is shared only through the male line, this means that women would also have inherited the same trait for longer life. In other words, because their fathers survive for as long as they do, daughters are dragged along for the ride. "Like nipples," explains Michael Gurven, an evolutionary anthropologist at the University of California, Santa Barbara. Men have nipples because women have nipples, even though they don't need them. Similarly, goes the patriarch hypothesis, women enjoy long life spans even though they don't need them, because men do.

When they explored the patriarch hypothesis years later, Singh, Morton, and Stone believed that Frank Marlowe's line of thinking didn't fully explain how menopause could have emerged. Running computer models to simulate how humans might have evolved in our early history, they found that adding a few genetic mutations for infertility into the population didn't have much of an effect on everyone's fertility as time wore on. These mutations just died out. "Fertility and survival remained high into old age. There was no menopause," they said. But when they added the critical element of older men preferring to have sex with younger women into their simulations, female menopause did pop up.

It was evidence, they claimed, that the patriarch hypothesis, slightly tweaked, could explain menopause in women. Grandmothers may be hardworking, but in the end it just comes down to sexual attractiveness.

———————

Like Kristen Hawkes, Frank Marlowe had also studied Hadza hunter-gatherers at close quarters for many years. The difference was that he came up with a different explanation for human longevity and menopause. So how did two distinguished researchers studying exactly the same group of people come to two such conflicting theories?

Anthropologist Alyssa Crittenden at the University of Nevada, Las Vegas, who has worked closely with Frank Marlowe, believes part of the reason may be because he and Hawkes were with the Hadza at different times, with an interval between them of almost two decades. These communities, as fragile as they are as they interact with the rest of the world, may have changed how they lived even in that short time. The same way, for example, as the women of the Nanadukan Agta in the Philippines began to abandon hunting.

But there are other explanations. "Part of it may lie in the sex of the researcher," Crittenden admits. "Science is supposed to be objective," but it's possible that their sex affected how they collected their data, she adds.

Hawkes and Marlowe now have their own scientific camps, each with their own versions of the past, one favoring powerful old men and the other favoring grandmothers. "The one I'm betting on really does make grandmothering the key to this special characteristic of our longevity," states Hawkes. For the patriarch hypothesis to work, she explains, there would have to be at least a few old men alive in the beginning to make these patriarchies happen in the first place, and the fact that our primate cousins don't have any elderly chimps or apes among them raises the question of where on earth these older men might have come from in large enough numbers. "The problem with his patriarch hypothesis is he has to somehow get to the place he wants to start," she tells me.

By the time I call Marlowe in 2015 he has Alzheimer's disease and isn't available for an interview. Alyssa Crittenden tells me that although she still very much respects his research, his scientific article on patriarchs hasn't stood the test of time as well as some of his other work. It has been cited far less by other researchers than Hawkes's papers on the grandmother hypothesis, for example.

Others, however, continue to disagree. When I ask evolutionary anthropologist Michael Gurven about the grandmother hypothesis, he is skeptical of it. In 2007 Gurven, together with Stanford University biology professor Shripad Tuljapurkar and Cedric Puleston, then a doctoral candidate at Stanford, published a paper titled "Why Men Matter: Mating Patterns Drive Evolution of Human Lifespan." In it they argue, along the lines of Frank Marlowe's patriarch hypothesis, that the general pattern for husbands to be older than their wives, along with a small number of high-status older men managing to mate with younger women, could partly account for why humans live so long.

Their view is that even if the grandmother hypothesis was true, men must have had a role to play as well. "You can't correctly estimate the force of selection if you leave men out of the picture," Puleston told a reporter from the *Stanford News Service* when their paper came out. "As a man myself, it's gratifying to know that men do matter."

Gurven these days takes the middle ground between men and women by suggesting that grandparents of both genders, not just the female ones, are responsible for our long life spans. He doesn't believe that women alone could account for such an important feature of human evolution. This two-sex model claims that it isn't only babysitting or food production that makes older people useful. Passing down knowledge from generation to generation could be one benefit, according to Gurven. Another could be in mediating conflict. Big-brained and complicated as humans are, skill usually comes with age. This wisdom sharing is something that both men and women could have played a role in through history.

The problem for everyone in this field is that the data are both scarce and messy. We can't know for sure how people lived many millennia ago. The Hadza, remarkable a window on the past though they might be, are nevertheless a small and dusty one. And evidence from other hunter-gatherer communities around the world is even more sketchy. This leaves room for speculation. Gurven is on the softer end of opposition to the grandmother hypothesis. Marlowe, Morton, Stone, and Singh have been on the harder end. But the trend here isn't difficult to spot: countertheories to the grandmother hypothesis appear to come mainly from men.

Gurven laughs when I ask him if there might even be some bias at work in his field of research. "You mean humans studying humans have bias?" he asks sarcastically. The rainbow of explanations for why humans live as long as they do and what makes older people useful in different societies means that many more things are possible than would have actually happened, he explains. It is this room for uncertainty that makes menopause such a volatile subject. From patriarchs to grandmothers, we may never know for sure who's right. "If you polled a whole bunch of people and asked them what they believe, would more women choose the grandmother hypothesis and more men the patriarch hypothesis? I wouldn't be surprised. . . . It's hard to remain completely unbiased," Gurven admits.

His opinion is that Morton, Stone, and Singh's hypothesis about men alone causing menopause is a case of wishful thinking. But he also believes that Kristen Hawkes has fought too hard for the grandmother hypothesis, even neglecting critiques of her evidence. It survives, he says, because it is sexy, not because it is right. "By throwing men under the bus, it seemed to be a radical new idea and people clung to that," he tells me.

Donna Holmes, an expert on the biology of aging based at the University of Idaho, agrees with Gurven on this point. She tells me that she has clashed with Hawkes over the grandmother hypothesis, and that she's still not convinced by it. "It was provocative and fresh. It made feminists happy, because it was grandmother friendly and went against the idea that older women are not valuable. It made liberals happy, because they like to think that aging can be 'natural' and accomplished without intervention by the evil pharmaceutical industry," says Holmes. "So it became very fashionable."

Alyssa Crittenden doesn't see it this way. "It's important to highlight the role that Kristen Hawkes played," she says. Torn by what she sees as compelling arguments both ways, she tells me, "Gun to my head, I choose the grandmother hypothesis." Over the many years since both hypotheses were originally published, the data have strengthened the grandmother hypothesis more than they have Frank Marlowe's patriarch hypothesis, she adds. "I'm continually blown away by the economic effort that postmenopausal women expend. . . . I really believe grandmothers have a really special role."

More than three decades since publishing her original paper on hardworking Hadza grandmothers, Kristen Hawkes maintains that the weight of evidence is on her side. "I had no notion that what old ladies were doing was going to turn out to be so important," she tells me. "It really highlights the extremely important effects that postfertile females have had on the direction of evolution in our lineage."

However controversial it might be, her research has helped bring older women into the evolutionary frame. A door has opened to a completely different and more positive way of thinking about aging. And today it sits inside a wider body of work questioning whether menopause should in fact be welcomed rather than feared. As far back as the 1970s American anthropologist Marcha Flint studied communities in Rajasthan in India, where women saw old age very differently. They told her it was a good thing, giving them a new social standing in their communities and more equality with men. Flint described negative American attitudes to menopause, in contrast, as a "syndrome." When menopause is seen as a curse rather than a blessing, women feel naturally less happy about it and also seem to report more symptoms.

This observation has been more recently supported by others. Researcher Beverley Ayers, when working in the psychology department at King's College London in 2011, argued that the way the Western medical profession has treated menopausal women has made them believe that menopause has more symptoms than it really does. In an article published in the *Psychologist* she and her colleagues explain that Western women have reported experiencing "hot flushes, night sweats, irregular and heavy periods, depression, headaches, insomnia, anxiety and weight gain," while in India, China, and Japan, these symptoms aren't nearly as common. One explanation might be that women are lumping in the effects of growing older with their experiences of menopause. If science tells them that menopause is a disease, they start feeling as though it is.

The story of menopause is the story of how science has sometimes failed women. But, as the grandmother hypothesis shows, science has provided alternative narratives, too, ones that not only challenge old preconceptions and tired stereotypes but also can be truly empowering. Indeed, Kristen Hawkes's latest work suggests that hardworking grandmothers may have appeared very early in human development, around two million years ago, meaning they could hold much more than just the key to human longevity. "It may have been helpful grandmothering that allowed the spread of genus

Homo out of Africa and into previously unoccupied regions of the temperate and tropical Old World," she speculates. In her version of the story of us, ancient grandmothers weren't just powerhouses in their families but vehicles for enormous change as humans migrated across the globe, tens of thousands of years ago. Age was no barrier to exercising their strength.

With the hard work of these women, everything was possible.

' AFTERWORD

*The feminists had destroyed the old image of woman, but they
could not erase the hostility, the prejudice, the discrimination
that still remained.*
—Betty Friedan, *The Feminine Mystique*, 1963

A science book on the shelves of the Wellcome Library in Bloomsbury, London, not too far from where I live, caught my eye. There among the rows of
academic journals and medical textbooks, tucked away in one corner, was a
small volume published in 1952 and titled *The Natural Superiority of Women*.
"The natural superiority of women is a biological fact, and a socially
overlooked piece of knowledge," wrote the author, a British American anthropologist by the name of Ashley Montagu. This bald statement sounded
radical to me when I first read it, but I could only imagine how much more
radical it must have sounded back in the 1950s, when women had the vote
but not much else. By the time I found his book, I had already pored over
many hundreds of pages of scientific literature stretching over two centuries
dedicated to the idea that women are somehow inferior to men. This little
volume was a rare exception. And it was written by a man. I bought my own
secondhand copy.

As I learned later, this wasn't Montagu's only controversial piece of
work. He was a prolific author who had lectured at Princeton and became
something of an intellectual celebrity in the postwar years, appearing on
American chat shows. When Hitler was committing atrocities against Jews

in Europe, he wrote about the fallacy of the biological idea of race. In his writings on women, he compared their subjugation to the historic treatment of black people in the United States. He campaigned against genital mutilation long before it was the high-profile issue it is today.

Montagu wasn't always Montagu. He was born Israel Ehrenberg to Jewish Russian immigrants in 1905 in east London—an upbringing that would have almost certainly made him a victim of anti-Semitism. Maybe that's why he ended up changing his name. He picked the eighteenth-century writer and feminist Lady Mary Wortley Montagu. She had been known for her travel writing from the Ottoman Empire and for advocating in favor of smallpox inoculation after she saw it being used effectively in Turkey. She was so sure this medical practice would save lives that she had her own children inoculated, long before it became common in England.

I don't know whether Lady Mary was any more of an inspiration to him beyond her name, but she seems to have been. In the pages of his book, Montagu looks at the biological measures by which we assume women are inferior to men. He uses data to show that, intellectually and physically, women aren't weak and feeble. And he makes a passionate case for improving the status of women. It's not always objective. In fact, at moments, he seems a little amused by his own idea. "If I sometimes poke a little light-hearted fun at my own sex, I hope that no man will be humorless enough to think that I am casting aspersions upon him," he reassures.

But Montagu is also clear that men have everything to gain from embracing change. He calls for flexible working patterns, in which parents can split child care evenly between them so both can enjoy the benefits of raising their kids. He asks husbands not to leave housework to stay-at-home wives, however much they dislike it. "Man is himself a problem in search of a solution," he writes. "When men understand that the best way to solve their own problem is to help women solve those that men have created for women, they will have taken one of the first significant steps toward its solution. . . . The truth will make men free as well as women." It's a message as timely then as it is now.

At this point, though, let me tell you the story of another anthropologist.

In 2015 Melvin Konner, who's based at Emory University, took inspiration from Montagu's book and wrote his own, titled *Women After All: Sex, Evolution, and the End of Male Supremacy*. He argues that some qualities common to women make them natural leaders in the modern age. "I happen to think it's superior to be less violent," Konner tells me, when I interview him about it. If brute strength is a large part of the reason for male supremacy, then in an age when strength matters less and violence appears to be declining, he says, women should naturally ascend. "I think it'll be a better world if women have more influence."

It doesn't sound all that radical now. After all, change is already underway. We have women leaders. Indeed, some critics have found Konner's arguments more than a little patronizing. But the simple idea of women being in charge, which may have been amusingly provocative when Montagu's *The Natural Superiority of Women* appeared on the shelves in 1952, is taken very differently these days. When Konner's book was serialized in the *Wall Street Journal*, within forty-eight hours he had more than seven hundred comments, many of them from a "men's rights movement." "There were some comments that were brief, but started and ended with 'fuck you,'" he recalls. Another told him, "There's no describing your kind of stupid." The response came as a shock. His wife had to double lock their doors.

The idea of women gaining power, Konner admits, "is threatening" to some.

That shouldn't really surprise us. When suffragists in the nineteenth and early twentieth centuries fought for the right to vote, they faced enormous opposition. It was a bitter, bloody battle. Thousands were imprisoned and some were tortured. Every wave for change in women's lives has brought with it the same kind of resistance.

And today, as women across the world fight for more freedoms and equality, there are again violent efforts to hold them back. According to the Guttmacher Institute, a research organization that aims to advance reproductive rights, the last five years have seen a sharp rise in attempts by some US states to impose restrictions on a woman's right to an abortion. Some of these are limits on abortion medication, others on private insurance coverage and rules about abortion clinics. "The sustained assault on abortion access is showing no signs of abating," a news release by the institute warned in January 2016.

Similarly, despite enormous efforts to raise awareness, female feticide in South Asia and female genital mutilation in Africa remain endemic. The spread of religious fundamentalism, which emphasizes female modesty, is also seeing the promise of female sexual freedom decaying right before our eyes.

A phenomenon known as the "Nordic Paradox" shows that equality under the law doesn't always guarantee women will be treated better. Iceland has among the highest levels of female participation in the labor market anywhere in the world, with heavily subsidized child care and equal parental leave for mothers and fathers. In Norway, since 2006, the law has required that at least 40 percent of listed company board members are women. Yet a report in May 2016 published in *Social Science and Medicine* reveals that Nordic countries have a disproportionately high rate of intimate partner violence against women. One theory to explain the paradox is that Nordic countries may be experiencing a backlash effect as traditional ideas of manhood and womanhood are challenged.

The world may seem better for women than it was in 1952, when Ashley Montagu wrote *The Natural Superiority of Women*, but in some ways it's worse. Resistance from certain corners is so powerfully toxic that it threatens to overturn the progress that's been made.

You may think these struggles have nothing to do with the lofty world of science. Academics often balk at the thought of mixing their work with politics. But when it comes to women, there's no avoiding it. Without taking into account how deeply unfair science has been to women in the past (and in some quarters, still is), it's impossible to be fairer in the future. And this is important for all of us. Because what science tells us about women profoundly shapes how society thinks about the sexes. The battle for minds in the fight for equality has to include the biological facts.

Pretty much every scientist I interviewed for this book who is working to challenge negative research about women told me she or he is a feminist. This doesn't make them any less brilliant at their work. In some cases, just the opposite. The social psychologist Carol Tavris, author of *The Mismeasure of Woman: Why Women Are Not the Better Sex, the Inferior Sex, or the Opposite Sex*, puts it to me this way: "Here is feminism, which involves ideological, political, and moral beliefs and goals. And here is science, which requires us to put our beliefs and assumptions, including those inspired by feminism, to empirical test. . . . For decades feminism has been a lens that illuminated biases in science. It made science better. Women began studying questions about women's lives—menstruation, pregnancy and childbirth, sexuality, work and careers, love—that most male researchers simply weren't interested in. When men did include women in their studies and found gender differences, they often concluded that women weren't just different from men, but deficient. So feminism was a crucial way to explode beliefs that people held that were just wrong."

When I set out to write this book, I wanted to get to the heart of the facts, even if they were uncomfortable. Where the facts weren't clear, I wanted to highlight the debates around them. I didn't want to show that one sex is inferior and the other superior (I don't believe that's a distinction anyone can even reasonably make). I just wanted to better understand the biological story about myself and other women. As I've learned, science is far from perfect. That's not the fault of the method but of us. We imperfect creatures crash its home and dirty its carpets with our feet. We throw our weight around when we should instead be its respectful guests. With us in charge, science can only be a self-correcting journey toward the truth. As such, none of the research I've written about represents the end of the story. Theories are only theories, waiting for more evidence.

But however unclear the research is in some areas, I did find reassurance that science has everything to offer women and men who want to live in a fairer world. Feminism can be a friend to science. It not only improves how science is done by pushing researchers to include the female perspective, but science in turn can also show us that we're not as different as we seem. Research to date suggests that humans survived, thrived, and spread across the globe through the efforts of everyone equally sharing the same work and responsibilities. For most of our history, we lived hand in hand. And our biology reflects this.

In some ways, of course, our biology makes no difference to how we live today. We've entered the epoch that scientists call the "anthropocene," in which humans are recognized to have had a profound impact on global ecosystems. We control our environments in ways that no other animal can. What's more, we control ourselves. We have birth control that can stop women getting pregnant and paternity tests that allow fathers to identify their children. Within decades, it may be possible to delay menopause far into old age. Artificial intelligence may eventually rewrite the laws of work and love. The world in which we evolved into humans isn't the same anymore. We've given ourselves the option to live any way we want.

In this world, then, it may seem strange that we're laboring under the same old stereotypes that have been around for centuries, that we're taking so long to make sexual equality a reality when the power to do it is entirely in our own hands. The cloudy window of the past has so distorted how we see society that we find it hard to imagine it another way. This is why science matters for every one of us. The job ahead for researchers is to keep cleaning the window until we see ourselves as we truly are, the way Ashley Montagu tried to do, and as so many pioneering researchers have done and continue to do today.

The facts are what will empower us to transform society for the better, into one that treats us as equals. Not just because this makes us civilized but because, as the evidence already shows, this makes us human.

ACKNOWLEDGMENTS

In spring 2014, Ian Tucker, an editor at the *Observer*, asked me to write a story about menopause. That story opened me up to a huge wealth of research on women, and in particular the controversies inside the scientific community about how to define what a woman is. It was also the kernel for this book. I'd like to thank my editor Louise Haines at Fourth Estate for taking a punt on *another* book on sex and gender, and for her invaluable guidance and ideas. Thanks also to Amy Caldwell at Beacon Press, who came up with the title. My agents, Peter Tallack and Tisse Takagi, provided enormous help in shaping the idea in the first place and improving the manuscript when it was finished.

My heartfelt thanks also to the Society of Authors and the K Blundell Trust, which generously provided me with a research grant to allow me the time to write, buy books, and travel for research. Without their help, as a working mother I simply wouldn't have been able to finish this. I hope their kindness to other writers in the same position as me never ends.

Many thanks to the Manuscripts Room at the Cambridge University Library for allowing me special access to Charles Darwin's private correspondence, including his letters to and from Caroline Kennard of Brookline, Massachusetts. Also, thanks to the Wellcome Library in London for access to their archive collection of pharmaceutical advertisements. The UK Intersex Association and Intersex UK gave me help and advice on issues surrounding intersex conditions.

I am also grateful to a number of friends and academics for their assistance in proofreading certain chapters. They include Richard Quinton, Norman Fenton, Paul Matthews, Tom Vulliamy, Jahnavi Phalkey, Denise

Sheer, Tim Power, Monica Niermann, Rainer Niermann, Rima Saini, and Mukul Devichand. Sarah Hrdy, Patricia Gowaty, and Robert Trivers were particularly generous with their time in answering my incessant questions. Dawn Starin was kind enough to offer her immense expertise in all my chapters on evolution. And Preeti Jha and Pramod Devichand read the whole thing with care and gave me invaluable feedback. But my deepest and most heartfelt thanks go to Peter Wrobel. I already thought he was the sharpest reader I knew, but his thoughts and fact checking of my manuscript have made me admire him more.

It's impossible to write a book and raise a two-year-old without the help of a village. The people I must thank most are my mother-in-law, Neena, who took considerable time off from her job as a doctor every week to look after her grandson, and my wonderful husband, Mukul, who was willing to forgo my company and to care for our son alone in the evenings and weekends so I could write and travel.

I am grateful to all my family and friends, as I am to my much-loved son, Aneurin, for making me smile every time I looked up from my reading. I hope he will one day read this book, because I wrote it with his future in mind.

REFERENCES

Introduction

Women's Engineering Society. "Statistics on Women in Engineering." Revised March 2016. http://www.wes.org.uk/sites/default/files/Women%20in%20Engineering %20Statistics%20March2016.pdf.

WISE. "Woman in the STEM Workforce." September 7, 2015. https://www.wise campaign.org.uk/resources/2015/09/women-in-the-stem-workforce.

National Science Foundation. "Women, Minorities, and Persons with Disabilities in Science and Engineering: 2015." http://www.nsf.gov/statistics/2015/nsf15311 /digest/nsf15311-digest.pdf.

Summers, Lawrence H. "Remarks at NBER Conference on Diversifying the Science and Engineering Workforce." Harvard University website, January 14, 2005. http:// www.harvard.edu/president/speeches/summers_2005/nber.php.

Hemel, Daniel J. "Summers' Comments on Women and Science Draw Ire." *Harvard Crimson*, January 14, 2005. http://www.thecrimson.com/article/2005/1/14/summers -comments-on-women-and-science/.

United Nations Educational, Scientific and Cultural Organization. "Women in Science." November 17, 2015. http://www.uis.unesco.org/ScienceTechnology/Pages /women-in-science-leaky-pipeline-data-viz.aspx.

Wolfinger, Nicholas. "For Female Scientists, There's No Good Time to Have Children." *Atlantic*, July 29, 2013. http://www.theatlantic.com/sexes/archive/2013/07 /for-female-scientists-theres-no-good-time-to-have-children/278165/.

US Bureau of Labor Statistics. "American Time Use Survey Summary." June 24, 2015. http://www.bls.gov/news.release/atus.nro.htm.

Institute for Women's Policy Research. "Pay Equity and Discrimination." http://www .iwpr.org/initiatives/pay-equity-and-discrimination.

UK Office for National Statistics, "Annual Survey of Hours and Earnings: 2016 Provisional Results." http://www.ons.gov.uk/employmentandlabourmarket/people inwork/earningsandworkinghours/bulletins/annualsurveyofhoursandearnings /2016provisionalresults#gender-pay-differences.

Moss-Racusin, Corinne A., et al. "Science Faculty's Subtle Gender Biases Favor Male Students." *Proceedings of the National Academy of Sciences USA* 109, no. 41 (2012): 16474–79.

Grunspan, Daniel Z., et al. "Males Under-Estimate Academic Performance of Their Female Peers in Undergraduate Biology Classrooms." *PLOS ONE* 11, no. 2 (2016).

Pattinson, Damian. *PLOS ONE Update on Peer Review Process* (blog). May 1, 2015. http:// blogs.plos.org/everyone/2015/05/01/plos-one-update-peer-review-investigation/.

Ghorayshi, Azeen. "'He Thinks He's Untouchable.'" *BuzzFeed News*, June 29, 2016. https://www.buzzfeed.com/azeenghorayshi/michael-katze-investigation?utm_term= .ctYdgQ3Zk#.rjvYQW9Z2.

Levin, Sam. "UC Berkeley Sexual Harassment Scandal Deepens amid Campus Pro-tests." *Guardian*, April 11, 2016. http://www.theguardian.com/us-news/2016/apr /11/uc-berkeley-sexual-harassment-scandal-protests.

Ghorayshi, Azeen. "Famous Berkeley Astronomer Violated Sexual Harassment Policies Over Many Years." *BuzzFeed News*, October 9, 2015. https://www.buzzfeed.com /azeenghorayshi/famous-astronomer-allegedly-sexually-harassed-students?utm _term=.ebnARDmVk#.lnMP9Og7b.

Mervis, Jeffrey. "Caltech Suspends Professor for Harassment." *Science*, January 12, 2016. http://www.sciencemag.org/news/2016/01/caltech-suspends-professor-harassment-0.

Harmon, Amy. "Chicago Professor Resigns amid Sexual Misconduct Investigation." *New York Times*, February 2, 2016. http://nyti.ms/207CCN0.

The Press Association. "Gender Gap in UK Degree Subjects Doubles in Eight Years, Ucas Study Finds." *Guardian*, January 5, 2016. https://www.theguardian.com/education /2016/jan/05/gender-gap-uk-degree-subjects-doubles-eight-years-ucas-study.

Schiebinger, Londa. *The Mind Has No Sex? Women in the Origins of Modern Science*. Cam-bridge, MA: Harvard University Press, 1989.

———. "Skeletons in the Closet: The First Illustrations of the Female Skeleton in Eighteenth-Century Anatomy." *Representations*, no. 14 (1986): 42–82.

Hamlin, Kimberly A. *From Eve to Evolution: Darwin, Science, and Women's Rights in Gilded Age America*. Chicago: University of Chicago Press, 2014.

Lucibella, Michael. "March 23, 1882: Birth of Emmy Noether," This Month in Physics History. *American Physical Society News* 22, no. 3 (March 2013).

Einstein, Albert. "Professor Einstein Writes in Appreciation of a Fellow-Mathematician." *New York Times*, May 5, 1935. http://www-groups.dcs.st-and.ac.uk /history/Obits2/Noether_Emmy_Einstein.html.

Carothers, Bobbi, and Harry Reis. "The Tangle of the Sexes." *New York Times*, April 20, 2013. http://nyti.ms/1A15mdP.

Ruti, Mari. *The Age of Scientific Sexism: How Evolutionary Psychology Promotes Gender Pro-filing and Fans the Battle of the Sexes*. New York: Bloomsbury Press, 2015.

Chapter 1: Woman's Inferiority to Man

Darwin, Charles. *The Descent of Man: Selection in Relation to Sex*. London: John Murray, 1871.

Schiebinger, Londa. *The Mind Has No Sex? Women in the Origins of Modern Science*. Cam-bridge, MA: Harvard University Press, 1989.

Hoeveler, J. David. *The Evolutionists: American Thinkers Confront Charles Darwin, 1860–1920*. Lanham, MD: Rowman & Littlefield, 2007.

Romanes, George John. "Mental Differences of Men and Women." *Popular Science Monthly* 31 (July 1887): 383–401.

Geddes, Patrick, and J. Arthur Thomson. *The Evolution of Sex*. Orig. 1889. New York: Scribner & Welford, 1890.

Hamlin, Kimberly A. *From Eve to Evolution: Darwin, Science, and Women's Rights in Gilded Age America*. Chicago: University of Chicago Press, 2014.

Egan, Maureen L. "Evolutionary Theory in the Social Philosophy of Charlotte Perkins Gilman." *Hypatia* 4, no. 1 (1989): 102–19.

Gamble, Eliza Burt. *The Evolution of Woman, an Inquiry into the Dogma of Her Inferiority to Man*. New York: Knickerbocker Press, 1894.

Baca, Katherine Ana Ericksen. "Eliza Burt Gamble and the Proto-Feminist Engagements with Evolutionary Theory." Undergraduate thesis, Harvard University, 2011.

Wolfe, A. B. "Sex Antagonism, by Walter Heape." Reviewed work. *American Journal of Sociology* 20, no. 4 (1915): 551.

Oudshoorn, Nelly. "Endocrinologists and the Conceptualization of Sex, 1920–1940." *Journal of the History of Biology* 23, no. 2 (1990): 163–86.

Van Den Wijngaard, Marianne. *Reinventing the Sexes: The Biomedical Construction of Femininity and Masculinity*. Bloomington: Indiana University Press, 1997.

Angier, Natalie. *Woman: An Intimate Geography*. London: Virago, 1999.

Wass, John. "The Fantastical World of Hormones." *Endocrinologist* (Spring 2015): 6–7.

Seymour, Jane Katherine. "The Medical Meanings of Sex Hormones: Clinical Uses and Concepts in *The Lancet*, 1929–1939." Dissertation, Wellcome Centre for the History of Medicine at University College London, 2005.

Fausto-Sterling, Anne. *Sexing the Body: Gender Politics and the Construction of Sexuality*. New York: Basic Books, 2000.

Zondek, Bernhard. "Mass Excretion of Oestrogenic Hormone in the Urine of the Stallion." *Nature* 133 (1934): 209–10.

Evans, Herbert M. "Endocrine Glands: Gonads, Pituitary, and Adrenals." *Annual Review of Physiology* 1 (1939): 577–652.

Sanday, Peggy Reeves. "Margaret Mead's View of Sex Roles in Her Own and Other Societies." *American Anthropologist* 82, no. 2 (1980): 340–48.

Coates, J. M., and J. Herbert. "Endogenous Steroids and Financial Risk Taking on a London Trading Floor." *Proceedings of the National Academy of Sciences USA* 105, no. 16 (2008): 6167–72.

Cueva, Carlos, et al. "Cortisol and Testosterone Increase Financial Risk Taking and May Destabilize Markets." *Scientific Reports* 5, no. 11206 (2015).

Chapter 2: Females Get Sicker but Males Die Quicker

Ramesh, Randeep. "Dozens of Female Babies' Body Parts Found in Disused Indian Well in New Delhi." *Guardian*, July 23, 2007.

Jha, Prabhat, et al. "Trends in Selective Abortions of Girls in India: Analysis of Nationally Representative Birth Histories from 1990 to 2005 and Census Data from 1991 to 2011." *Lancet* 377 (2011): 1921–28.

United Nations Population Fund. *Trends in Sex Ratio at Birth and Estimates of Girls Missing at Birth in India 2001–2008*. New Delhi: UNFPA, 2010.

United Nations. "Health." Chapter 2 in *The World's Women 2015: Trends and Statistics*. http://unstats.un.org/unsd/gender/downloads/WorldsWomen2015_chapter2_t.pdf.

John, Mary E. *Sex Ratios and Gender Biased Sex Selection: History, Debates and Future Directions.* UN Women, 2014.

Yamanaka, Miki, and Ann Ashworth. "Differential Workloads of Boys and Girls in Rural Nepal and Their Association with Growth." *American Journal of Human Biology* 14, no. 3 (2002): 356–63.

Lawn, Joy E., et al. "Beyond Newborn Survival: The World You Are Born into Determines Your Risk of Disability-Free Survival." *Pediatric Research* 74, no. S1 (2013): 1–3.

Peacock, Janet L., et al. "Neonatal and Infant Outcome in Boys and Girls Born Very Prematurely." *Pediatric Research* 71, no. 3 (2012): 305–10

Buckberry, Sam, et al. "Integrative Transcriptome Meta-Analysis Reveals Widespread Sex-Biased Gene Expression at the Human Fetal–Maternal Interface." *Molecular Human Reproduction* 20, no. 8 (2014): 810–19.

Austad, Steven N. "Why Women Live Longer Than Men: Sex Differences in Longevity." *Gender Medicine* 3, no. 2 (2006): 79–92.

Austad, Steven N., and Andrzej Bartke. "Sex Differences in Longevity and in Responses to Anti-Aging Interventions: A Mini-Review." *Gerontology* 62, no. 1 (2016): 40–6.

Hitchman, Sara C., and Geoffrey T. Fong. "Gender Empowerment and Female-to-Male Smoking Prevalence Ratios." *Bulletin of the World Health Organization* 89, no. 3 (2011): 161–240.

"Numbers of Living Supercentenarians as of Last Update." Gerontology Research Group. Last updated July 9, 2016. http://www.grg.org/Adams/TableE.html.

Oertelt-Prigione, Sabine. "The Influence of Sex and Gender on the Immune Response." *Autoimmunity Reviews* 11, no. 6 (2012): A479–85.

Robinson, D. P., and S. L. Klein. "Pregnancy and Pregnancy-Associated Hormones Alter Immune Responses and Disease Pathogenesis." *Hormones and Behavior* 62, no. 3 (2012): 263–71.

Giefing-Kroll, Carmen, et al. "How Sex and Age Affect Immune Responses, Susceptibility to Infections, and Response to Vaccination." *Aging Cell* 14, no. 3 (2015): 309–21.

Furman, David, et al. "Systems Analysis of Sex Differences Reveals an Immunosuppressive Role for Testosterone in the Response to Influenza Vaccination." *Proceedings of the National Academy of Sciences* 11, no. 2 (2014): 869–74.

Ngo, S. T., F. J. Steyn, and P. A. McCombe. "Gender Differences in Autoimmune Disease." *Frontiers in Neuroendocrinology* 35, no. 3 (2014): 347–69.

Fairweather, DeLisa, Sylvia Frisancho-Kiss, and Noel R. Rose. "Sex Differences in Autoimmune Disease from a Pathological Perspective." *American Journal of Pathology* 173, no. 3 (2008): 600–9.

Maher, Brendan. "Women Are More Vulnerable to Infections." *Nature News*, July 26, 2013. http://www.nature.com/news/women-are-more-vulnerable-to-infections-1.13456.

Goldhill, Olivia. "Period Pain Can Be 'Almost as Bad as a Heart Attack.' Why Aren't We Researching How to Treat It?" *Quartz*, February 15, 2016. http://qz.com/611774/period-pain-can-be-as-bad-as-a-heart-attack-so-why-arent-we-researching-how-to-treat-it/.

Din, Nafees U., et al. "Age and Gender Variations in Cancer Diagnostic Intervals in 15 Cancers: Analysis of Data from the UK Clinical Practice Research Datalink." *PLOS ONE* 10, no. 5 (2015).

Ropers, H. H., and B. C. Hamel. "X-linked Mental Retardation." *Nature Reviews Genetics* 6, no. 1 (2005): 46–57.

Arnold, Arthur P., et al. "The Importance of Having Two X Chromosomes." *Philosophical Transactions of the Royal Society B* 371, no. 1688 (2016).

Berletch, Joel B., et al. "Genes That Escape from X Inactivation." *Human Genetics* 130, no. 2 (2011): 237–45.

Prothero, Katie E., Jill M. Stahl, and Laura Carrel. "Dosage Compensation and Gene Expression on the Mammalian X Chromosome: One Plus One Does Not Always Equal Two." *Chromosome Research* 17, no. 5 (2009): 637–48.

Richardson, Sarah S. *Sex Itself: The Search for Male and Female in the Human Genome.* Chicago: University of Chicago Press, 2013.

Beery, Annaliese, and Irving Zucker. "Sex Bias in Neuroscience and Biomedical Research." *Neuroscience and Biobehavioral Reviews* 35, no. 3 (2011): 565–72.

Ah-King, Malin, Andrew B. Barron, and Marie E. Herberstein. "Genital Evolution: Why Are Females Still Understudied?" *PLOS Biology* 12, no. 5 (2014).

Institute of Medicine. *Women's Health Research: Progress, Pitfalls, and Promise.* Washington, DC: National Academies Press, 2010.

Fadiran, Emmanuel O., and Lei Zhang. "Effects of Sex Differences in the Pharmacokinetics of Drugs and Their Impact on the Safety of Medicines in Women." In *Medicines for Women*, edited by Mira Harrison-Woolrych, 41–68. Auckland, NZ: ADIS, 2015.

Heinrich, Janet. "Drugs Withdrawn from Market," GAO-01–286R. US Government Accountability Office. January 19, 2001. http://www.gao.gov/new.items/d01286r.pdf.

"Exclusion from Clinical Trials Harming Women's Health." European Commission Community Research and Development Information Service. Last updated March 8, 2007. http://cordis.europa.eu/news/rcn/27270_en.html.

Digitalis Investigation Group. "The Effect of Digoxin on Mortality and Morbidity in Patients with Heart Failure." *New England Journal of Medicine* 336, no. 8 (1997): 525–33.

Rathore, S. S., Y. Wang, and H. M. Krumholz. "Sex-Based Differences in the Effect of Digoxin for the Treatment of Heart Failure." *New England Journal of Medicine* 347, no. 18 (2002): 1403–11.

Flory, J. H., et al. "Observational Cohort Study of the Safety of Digoxin Use in Women with Heart Failure." *British Medical Journal Open*, April 13, 2012. http://bmjopen.bmj.com/content/2/2/e000888.full#ref-1.

Greenblatt, D. J., et al. "Gender Differences in Pharmacokinetics and Pharmacodynamics of Zolpidem Following Sublingual Administration." *Journal of Clinical Pharmacology* 54, no. 3 (2014): 282–90.

Richardson, Sarah S., et al. "Focus on Preclinical Sex Differences Will Not Address Women's and Men's Health Disparities." Opinion. *Proceedings of the National Academy of Sciences United States of America* 112, no. 44 (2015): 13419–20.

Richardson, Sarah S. "Is the New NIH Policy Good for Women?" *Catalyst: Feminism, Theory, and Technoscience* 1, no. 1 (2015).

Clayton, Janine A., and Francis S. Collins. "NIH to Balance Sex in Cell and Animal Studies," Policy. *Nature* 509, no. 7500 (2014): 282–83.

Chapter 3: A Difference at Birth

Martin, Carol Lynn, and Diane Ruble. "Children's Search for Gender Cues: Cognitive Perspectives on Gender Development." *Current Directions in Psychological Science* 13, no. 2 (2004): 67–70.

Eliot, Lise. *Pink Brain, Blue Brain: How Small Differences Grow into Troublesome Gaps— And What We Can Do About It*. Boston: Houghton Mifflin Harcourt, 2009.

Connellan, Jennifer, et al. "Sex Differences in Human Neonatal Social Perception." *Infant Behavior and Development* 23, no. 1 (2000): 113–18.

Baron-Cohen, Simon. "The Truth about Science and Sex." *Guardian*, January 27, 2005. http://www.theguardian.com/science/2005/jan/27/science.educationsgendergap.

Pinker, Steven, and Elizabeth Spelke. "The Science of Gender and Science: Pinker vs. Spelke: A Debate." Edge.org, May 16, 2005. http://edge.org/event/the-science -of-gender-and-science-pinker-vs-spelke-a-debate.

Cronin, Helena. "The Vital Statistics." *Guardian*, March 12, 2005. http://www.theguardian .com/world/2005/mar/12/gender.comment.

Larimore, Walt, and Barbara Larimore. *His Brain, Her Brain: How Divinely Designed Differences Can Strengthen Your Marriage*. Grand Rapids, MI: Zondervan, 2008.

Baron-Cohen, Simon. "The Extreme Male Brain Theory of Autism." *Trends in Cognitive Sciences* 6, no. 6 (2002): 248–54.

———. *The Essential Difference*. New York: Perseus Books, 2003.

Wolpert, Lewis. *Why Can't a Woman Be More Like a Man?* London: Faber & Faber, 2014.

Goy, Robert W., and Bruce S. McEwen. *Sexual Differentiation of the Brain: Based on a Work Session of the Neurosciences Research Program*. Cambridge, MA: MIT Press, 1980.

Kolata, Gina. "Math Genius May Have Hormonal Basis." *Science* 222, no. 4630 (1983): 1312.

Geschwind, Norman, and Albert M. Galaburda. *Cerebral Dominance: The Biological Foundations*. Cambridge, MA: Harvard University Press, 1984.

McManus, I. C., and M. P. Bryden. "Geschwind's Theory of Cerebral Lateralization: Developing a Formal, Causal Model." *Psychological Bulletin* 110, no. 2 (1991): 237–53.

Bryden, M. P., I. C. McManus, and M. B. Bulman-Fleming. "Evaluating the Empirical Support for the Geschwind-Behan-Galaburda Model of Cerebral Lateralization." *Brain and Cognition* 26, no. 2 (1994): 103–67.

Kolata, Gina. "Sex Hormones and Brain Development." *Science* 205, no. 4410 (1979): 985–87.

Van den Wijngaard, Marianne. "The Acceptance of Scientific Theories and Images of Masculinity and Femininity: 1959–1985." *Journal of the History of Biology* 24, no. 1 (1991): 19–49.

Hines, Melissa. "Sex-Related Variation in Human Behavior and the Brain." *Trends in Cognitive Sciences* 14, no. 10 (2010): 448–56.

Wallen, Kim, and Janice M. Hassett. "Sexual Differentiation of Behavior in Monkeys: Role of Prenatal Hormones." *Journal of Neuroendocrinology* 21, no. 4 (2009): 421–26.

Alexander, G. M., and M. Hines. "Sex Differences in Response to Children's Toys in Nonhuman Primates (*Cercopithecus aethiops sabaeus*)." *Evolution and Human Behavior* 23, no. 6 (2002): 467–79.

Hines, Melissa, et al. "Testosterone During Pregnancy and Gender Role Behavior of Preschool Children: A Longitudinal, Population Study." *Child Development* 73, no. 6 (2002): 1678–87.

Jadva, V., M. Hines, and S. Golombok. "Infants' Preferences for Toys, Colors, and Shapes: Sex Differences and Similarities." *Archives of Sexual Behavior* 39, no. 6 (2010): 1261–73.

Auyeung, Bonnie, et al. "Fetal Testosterone Predicts Sexually Differentiated Childhood Behavior in Girls and in Boys." *Psychological Science* 20, no. 2 (2009): 144–48.

Maccoby, Eleanor Emmons, and Carol Nagy Jacklin. *The Psychology of Sex Differences.* Palo Alto, CA: Stanford University Press, 1974.

Gurwitz, Sharon B. "*The Psychology of Sex Differences* by Eleanor Emmons Maccoby, Carol Nagy Jacklin." Reviewed work. *American Journal of Psychology* 88, no. 4 (1975): 700–703.

Hyde, Janet Shibley. "The Gender Similarities Hypothesis." *American Psychologist* 60, no. 6 (2005): 581–92.

Colom, Roberto. "Negligible Sex Differences in General Intelligence." *Intelligence* 28, no. 1 (2000): 57–68.

Johnson, Wendy, Andrew Carothers, and Ian J. Deary. "Sex Differences in Variability in General Intelligence: A New Look at the Old Question." *Perspectives on Psychological Science* 3, no. 6 (2008): 518–31.

Leslie, Sarah-Jane, et al. "Expectations of Brilliance Underlie Gender Distributions across Academic Disciplines." *Science* 347, no. 6219 (2015): 262–65.

Grossi, Giordana, and Alison Nash. "Picking Barbie's Brain: Inherent Sex Differences in Scientific Ability?" *Journal of Interdisciplinary Feminist Thought* 2, no. 1 (2007): article 5.

Fine, Cordelia. *Delusions of Gender: The Real Science Behind Sex Differences.* London: Icon Books, 2010.

Lutchmaya, Svetlana, Simon Baron-Cohen, and Peter Raggatt. "Foetal Testosterone and Eye Contact in 12-Month-Old Human Infants." *Infant Behavior and Development* 25, no. 3 (2002): 327–35.

Lombardo, Michael V., et al. "Fetal Testosterone Influences Sexually Dimorphic Gray Matter in the Human Brain." *Journal of Neuroscience* 32, no. 2 (2012): 674–80.

Baron-Cohen, Simon, et al. "Elevated Fetal Steroidogenic Activity in Autism." *Molecular Psychiatry* 20 (2014): 369–76.

Kung, Karson T. F., et al. "No Relationship Between Prenatal Androgen Exposure and Autistic Traits: Convergent Evidence from Studies of Children with Congenital Adrenal Hyperplasia and of Amniotic Testosterone Concentrations in Typically Developing Children." *Journal of Child Psychology and Psychiatry* (2016). Published online July 27, 2016.

Jordan-Young, Rebecca M. *Brain Storm: The Flaws in the Science of Sex Differences.* Cambridge, MA: Harvard University Press, 2010.

Davis, Shannon N., and Barbara J. Risman. "Feminists Wrestle with Testosterone: Hormones, Socialization and Cultural Interactionism as Predictors of Women's Gendered Selves." *Social Science Research* 49 (2015): 110–25.

Ruigroka, Amber N. V., et al. "A Meta-Analysis of Sex Differences in Human Brain Structure." *Neuroscience and Biobehavioral Reviews* 39 (2014): 34–50.

Chapter 4: The Missing Five Ounces of the Female Brain

Gardener, Helen H. *Facts and Fictions of Life*, Boston: Arena, 1893.

———. "Sex and Brain Weight." Letter to the editor. *Popular Science Monthly* 31, no. 10 (June 1887): 266–68.

Hammond, William. "Men's and Women's Brains." Letter to the editor. *Popular Science Monthly* 31, no. 28 (August 1887): 554–58.

Romanes, George John. "Mental Differences of Men and Women." *Popular Science Monthly* 31 (July 1887).

"Noted Suffragette's Brain as Good as a Man's Cornell Anatomist Finds, Disproving Old Theory." *Cornell Daily Sun*, September 29, 1927.

Grahame, Arthur. "Why You May Wear a Small Hat and Still Have a Big Mind." *Popular Science Monthly* (December 1926): 15–16.

Lecky, Prescott. "Are Women as Smart as Men?" *Popular Science Monthly* (July 1928): 28–29.

Gur, Ruben, et al. "Sex and Handedness Differences in Cerebral Blood Flow During Rest and Cognitive Activity." *Science* 217 (1982): 659–61.

"Men and Women: Are We Wired Differently?" *TODAY Health* (blog). December 14, 2006. http://www.today.com/id/16187129/ns/today-today_health/t/men-women -are-we-wired-differently/#.V2pWhWOhY3A.

Ingalhalikar, Madhura, et al. "Sex Differences in the Structural Connectome of the Human Brain." *Proceedings of the National Academy of Sciences of the United States of America* 111, no. 2 (2014): 823–28.

Gur, Ruben C., et al. "Sex Differences in Brain Gray and White Matter in Healthy Young Adults: Correlations with Cognitive Performance." *Journal of Neuroscience* 19, no. 10 (1999): 4065–72.

Khazan, Olga. "Male and Female Brains Really Are Built Differently." *Atlantic*, December 2, 2013. http://www.theatlantic.com/health/archive/2013/12/male-and-female -brains-really-are-built-differently/281962/.

Gray, Richard. "Brains of Men and Women Are Poles Apart." *Telegraph*, December 3, 2013. http://www.telegraph.co.uk/news/science/science-news/10491096/Brains-of -men-and-women-are-poles-apart.html.

Haines, Lester. "Women Crap at Parking: Official." *Register*, December 4, 2013. http:// www.theregister.co.uk/2013/12/04/brain_study_shocker/.

Gur, Ruben C., et al. "Age Group and Sex Differences in Performance on a Computerized Neurocognitive Battery in Children Age 8–21." *Neuropsychology* 26, no. 2 (2012): 251–65.

Sacher, Julia, et al. "Sexual Dimorphism in the Human Brain: Evidence from Neuroimaging." *Magnetic Resonance Imaging* 31 (2013): 366–75.

"Brain Connectivity Study Reveals Striking Differences Between Men and Women." News release. Perelman School of Medicine, University of Pennsylvania, December 2, 2013. http://www.uphs.upenn.edu/news/news_releases/2013/12/verma/.

Sample, Ian. "Male and Female Brains Wired Differently, Scans Reveal." *Guardian*, December 2, 2013. http://www.theguardian.com/science/2013/dec/02/men-women -brains-wired-differently.

Connor, Steve. "The Hardwired Difference Between Male and Female Brains Could Explain Why Men Are 'Better at Map Reading.'" *Independent*, December 3, 2013.

http://www.independent.co.uk/life-style/the-hardwired-difference-between-male
-and-female-brains-could-explain-why-men-are-better-at-map-8978248.html.

Bennett, Craig M., et al. "Neural Correlates of Interspecies Perspective Taking in the
Post-mortem Atlantic Salmon: An Argument for Multiple Comparisons Correc-
tion." *Journal of Serendipitous and Unexpected Results* 1, no. 1 (2009): 1–5.

Bennett, Craig M. "The Story Behind the Atlantic Salmon." *Prefrontal.org* (blog). September 18, 2009. http://prefrontal.org/blog/2009/09/the-story-behind-the
-atlantic-salmon/.

Button, Katherine S., et al. "Power Failure: Why Small Sample Size Undermines the
Reliability of Neuroscience." *Nature Reviews Neuroscience* 14 (2013): 365–76.

Rippon, Gina, et al. "Recommendations for Sex/Gender Neuroimaging Research: Key
Principles and Implications for Research Design, Analysis, and Interpretation."
Frontiers in Human Neuroscience 8, no. 650 (2014).

Fine, Cordelia. "Gender Differences Found in Brain Wiring: Insight or Neurosexism?"
Popular Science, December 5, 2013. http://www.popsci.com/article/gender-differences
-found-brain-wiring-insight-or-neurosexism.

Joel, Daphna, and Ricardo Tarrasch. "On the Mis-presentation and Misinterpretation
of Gender-Related Data: The Case of Ingalhalikar's Human Connectome Study."
Letter. *Proceedings of the National Academy of Sciences of the United States of America*
111, no. 6 (2014).

Joel, Daphna, and Ricardo Tarrasch. "On the Mis-Presentation and Misinterpretation
of Gender-Related Data: The Case of Ingalhalikar's Human Connectome Study."
PNAS 111, no. 6 (February 11, 2014), http://www.pnas.org/content/111/6/E637
.full?keytype2=tf_ipsecsha&ijkey=4183bcb77bcb8782c9324a9abf711223af7bbe9f.

Tan, Anh, et al. "The Human Hippocampus Is Not Sexually-Dimorphic: Meta-Analysis
of Structural MRI Volumes." *NeuroImage* 124 (2016): 350–66.

Cahill, Larry. "Equal ≠ The Same: Sex Differences in the Human Brain." *Cerebrum*,
April 2014.

———. "A Half-Truth Is a Whole Lie: On the Necessity of Investigating Sex Influences
on the Brain." *Endocrinology* 153, no. 6 (2012): 2541–43.

Short, Nigel. "Vive la Différence." *New in Chess*, February 2015. http://en.chessbase.com
/post/vive-la-diffrence-the-full-story.

Halpern, Diane F., et al. "Education Forum: The Pseudoscience of Single-Sex Schooling." *Science* 333 (2011): 1706–7.

O'Connor, Cliodhna, and Helene Joffe. "Gender on the Brain: A Case Study of Science
Communication in the New Media Environment." *PLOS ONE* 9, no. 10 (2014).

Maguire, Eleanor A., Katherine Woollett, and Hugo J. Spiers. "London Taxi Drivers
and Bus Drivers: A Structural MRI and Neuropsychological Analysis." *Hippocampus*
16 (2006): 1091–101.

May, Arne. "Experience-Dependent Structural Plasticity in the Adult Human Brain."
Trends in Cognitive Sciences 15, no. 10 (2011): 475–82.

Fine, Cordelia, et al. "Plasticity, Plasticity, Plasticity . . . and the Rigid Problem of Sex."
Trends in Cognitive Sciences 17, no. 11 (2013): 550–51.

Miller, David I., and Diane F. Halpern. "The New Science of Cognitive Sex Differences." *Trends in Cognitive Sciences* 18, no. 1 (2014): 37–45.

Joel, Daphna. "Male or Female? Brains Are Intersex." *Frontiers in Integrative Neuroscience* 5, no. 57 (2011).

Joel, Daphna, et al. "Sex beyond the Genitalia: The Human Brain Mosaic." *Proceedings of the National Academy of Sciences of the United States of America.* Published online November 30, 2015. http://www.pnas.org/content/early/2015/11/24/1509654112.

Shors, Tracey J., Chadrick Chua, and Jacqueline Falduto. "Sex Differences and Opposite Effects of Stress on Dendritic Spine Density in the Male versus Female Hippocampus." *Journal of Neuroscience* 21, no. 16 (2001): 6292–97.

Dubb, Abraham, et al. "Characterization of Sexual Dimorphism in the Human Corpus Callosum." *NeuroImage* 20 (2003): 512–19.

Chapter 5: Women's Work

Hrdy, Sarah Blaffer. *The Woman That Never Evolved.* Cambridge, MA: Harvard University Press, 1981.

————. *The Langurs of Abu: Female and Male Strategies of Reproduction.* Cambridge, MA: Harvard University Press, 1977.

Prüfer, Kay, et al. "The Bonobo Genome Compared with the Chimpanzee and Human Genomes." *Nature* 486 (2012): 527–31.

Hrdy, Sarah Blaffer. *Mothers and Others: The Evolutionary Origins of Mutual Understanding.* Cambridge, MA: Belknap Press of Harvard University Press, 2009.

Rosenberg, Karen, and Wenda R. Trevathan. "Birth, Obstetrics and Human Evolution." *BJOG: An International Journal of Obstetrics and Gynaecology* 109 (2002): 1199–206.

Hrdy, Sarah Blaffer. "The Past, Present, and Future of the Human Family." The Tanner Lectures on Human Values, University of Utah, February 27–28, 2001.

Magurran, Anne. "Maternal Instinct." *New York Times,* January 23, 2000. http://www .nytimes.com/2000/01/23/books/maternal-instinct.html?pagewanted=all.

Craig, Michael. "Perinatal Risk Factors for Neonaticide and Infant Homicide: Can We Identify Those at Risk?" *Journal of the Royal Society of Medicine* 97, no. 2 (2004): 57–61.

Bribiescas, Richard. *Men: Evolutionary and Life History.* Cambridge, MA: Harvard University Press, 2006.

Sear, Rebecca, and David A. Coall. "How Much Does Family Matter? Cooperative Breeding and the Demographic Transition." *Population and Development Review* 37 (2011): 81–112.

Scommegna, Paola. "More U.S. Children Raised by Grandparents." Population Reference Bureau, March 2012. http://www.prb.org/Publications/Articles/2012/US -children-grandparents.aspx.

Muller, Martin N., Frank W. Marlowe, Revocatus Bugumba, and Peter T. Ellison. "Testosterone and Paternal Care in East African Foragers and Pastoralists." *Proceedings of the Royal Society B* 276, no. 1655 (2009): 347–54.

Walker, Robert S., Mark V. Flinn, and Kim R. Hill. "Evolutionary History of Partible Paternity in Lowland South America." *Proceedings of the National Academy of Sciences* 107, no. 45 (2010): 19195–200.

Lee, Richard B., and Irven DeVore, eds. *Man the Hunter.* Chicago: Aldine, 1968.

Washburn, Sherwood, and Chet Lancaster. "The Evolution of Hunting." In *Man the Hunter,* edited by Richard B. Lee and Irven DeVore, 293–303. Chicago: Aldine, 1968.

Ardrey, Robert. *The Hunting Hypothesis: A Personal Conclusion Concerning the Evolutionary Nature of Man*. New York: Atheneum, 1976.

Slocum, Sally. "Woman the Gatherer: Male Bias in Anthropology." In *Toward an Anthropology of Women*, edited by Rayna R. Reiter, 36–50. New York: Monthly Review Press, 1975. Originally published under the name Sally Linton in 1971.

———. "Women as Shapers of the Human Adaptation." In *Woman the Gatherer*, edited by Frances Dahlberg, 75–120. New Haven, CT: Yale University Press, 1981.

Zihlman, Adrienne. "The Real Females of Human Evolution." *Evolutionary Anthropology* 21, no. 6 (2012): 270–76.

———. "Engendering Human Evolution." In *A Companion to Gender Prehistory*, edited by Diane Bolger. Chichester: Wiley-Blackwell, 2013.

O'Connell, James F., Kristen Hawkes, K. D. Lupo, and N. G. Blurton Jones. "Male Strategies and Plio-Pleistocene Archaeology." *Journal of Human Evolution* 43, no. 6 (2002): 831–72.

Hawkes, Kristen, James F. O'Connell, and James E. Coxworth. "Family Provisioning Is Not the Only Reason Men Hunt: A Comment on Gurven and Hill." *Current Anthropology* 51, no. 2 (2010): 259–64.

Gurven, Michael, and Kim Hill. "Why Do Men Hunt? A Reevaluation of 'Man the Hunter' and the Sexual Division of Labor." *Current Anthropology* 50, no. 1 (2009): 51–74.

Kaplan, Hillard S., Paul L. Hooper, and Michael Gurven. "The Evolutionary and Ecological Roots of Human Social Organization." *Philosophical Transactions of the Royal Society B: Biological Sciences* 364, no. 1533 (2009): 3289–99.

Piantadosi, Steven, and Celeste Kidd. "Extraordinary Intelligence and the Care of Infants." *Proceedings of the National Academy of Sciences Early Edition*, approved for publication March 30, 2016. http://www.pnas.org/content/early/2016/05/18/1506752113.abstract.

Zuk, Marlene. *Paleofantasy: What Evolution Really Tells Us About Sex, Diet, and How We Live*. New York: W. W. Norton, 2013.

O'Connor, Anahad. "A Marathon Runner Delivers a Baby." *New York Times*, October 11, 2011. http://well.blogs.nytimes.com/2011/10/11/a-marathon-runner-delivers-a-baby/?_r=0.

Estioko-Griffin, Agnes. "Women as Hunters: The Case of an Eastern Cagayan Agta Group." In *The Agta of Northeastern Luzon: Recent Studies*, edited by P. Bion Griffin and Agnes Estioko-Griffin. Cebu City, Philippines: University of San Carlos, 1985.

Estioko-Griffin, Agnes, and P. Bion Griffin. "Woman the Hunter: The Agta." In *Woman the Gatherer*, edited by Frances Dahlberg, 121–51. New Haven, CT: Yale University Press, 1981.

Goodman, Madeleine J., et al. "The Compatibility of Hunting and Mothering Among the Agta Hunter-Gatherers of the Philippines." *Sex Roles* 12, no. 11 (1985): 1199–209.

Dyble, Mark, et al. "Sex Equality Can Explain the Unique Social Structure of Hunter-Gatherer Bands." *Science* 348, no. 6236 (2015): 796–98.

Hill, Kim, et al. "Hunter-Gatherer Inter-band Interaction Rates: Implications for Cumulative Culture." *PLOS ONE* 9, no. 7 (2014): 1–9.

Bliege Bird, Rebecca. "Fishing and the Sexual Division of Labor Among the Meriam." *American Anthropologist* 109, no. 3 (2007): 442–51.

Bliege Bird, Rebecca, and Brian F. Codding. "The Sexual Division of Labor." In *Emerging Trends in the Social and Behavioral Sciences*, edited by R. A. Scott et al. Hoboken, NJ: John Wiley & Sons, May 15, 2015.

Hurtado, Ana Magdalena, et al. "Female Subsistence Strategies among Ache Hunter-Gatherers of Eastern Paraguay." *Human Ecology* 13, no. 1 (1985): 1–28.

Morbeck, Mary Ellen, Alison Galloway, and Adrienne L. Zihlman. *The Evolving Female: A Life History Perspective*. Princeton, NJ: Princeton University Press, 1997.

Chapter 6: Choosy, Not Chaste

Clark, Russell D., and Elaine Hatfield. "Gender Differences in Receptivity to Sexual Offers." *Journal of Psychology and Human Sexuality* 2, no. 1 (1989): 39–55.

———. "Love in the Afternoon." *Psychological Inquiry* 14, nos. 3 and 4 (2003): 227–31.

Bateman, Angus J. "Intra-Sexual Selection in Drosophila." *Heredity* 2 (1948): 349–68.

Trivers, Robert L., "Parental Investment and Sexual Selection." In *Sexual Selection and the Descent of Man*, edited by Bernard Campbell, 136–79. Chicago: Aldine, 1972.

Symons, Donald. *The Evolution of Human Sexuality*. New York: Oxford University Press, 1979.

Geertz, Clifford. "Sociosexology." *New York Review of Books*, January 24, 1980. http://www.nybooks.com/articles/1980/01/24/sociosexology/.

Buss, David M. *The Evolution of Desire: Strategies of Human Mating*. New York: Basic Books, 1994.

Pinker, Steven. "Boys Will Be Boys," Talk of the Town. *New Yorker*, February 9, 1998, 30–31.

———. *The Blank Slate: The Modern Denial of Human Nature*. New York: Viking, 2002.

———. "Sex Ed." *New Republic*, February 14, 2005. https://newrepublic.com/article/68044/sex-ed.

Gould, Stephen Jay. "Freudian Slip." *Natural History*, no. 96 (1987): 14–21.

Darwin, Charles. *The Descent of Man: Selection in Relation to Sex*. London: John Murray, 1871.

Miller, Geoffrey. *The Mating Mind: How Sexual Choice Shaped the Evolution of Human Nature*. London: Vintage, 2000.

Reich, Eugenie Samuel. "Symmetry Study Deemed a Fraud." *Nature*, May 3, 2013. http://www.nature.com/news/symmetry-study-deemed-a-fraud-1.12932.

Starin, Dawn. "She's Gotta Have It." *Africa Geographic* (May 2008): 57–62.

Tang-Martínez, Zuleyma. "Bateman's Principles: Original Experiment and Modern Data For and Against." In *Encyclopedia of Animal Behavior*, edited by M. D. Breed and J. Moore, 166–76. London: Elsevier/Academic Press, 2010.

Symons, Donald. "Another Woman That Never Existed." Review. *Quarterly Review of Biology* 57, no. 3 (1982): 297–300.

Hrdy, Sarah Blaffer. "Empathy, Polyandry, and the Myth of the Coy Female." In *Feminist Approaches to Science*, edited by Ruth Bleier, 119–46. New York: Pergamon Press, 1986.

———. "The Evolution of Human Sexuality: The Latest Word and the Last." *Quarterly Review of Biology* 54, no. 3 (1979): 309–14.

Bluhm, Cynthia, and Patricia Adair Gowaty. "Social Constraints on Female Mate Preferences in Mallards, *Anas platyrhynchos*, Decrease Offspring Viability and Mother Productivity." *Animal Behaviour* 68, no. 5 (2004): 977–83.

Milius, Susan. "If Mom Chooses Dad, More Ducklings Survive." *Science News* 156, no. 1 (1999): 6.

Drickamer, Lee C., Patricia Adair Gowaty, and Christopher M. Holmes. "Free Female Mate Choice in House Mice Affects Reproductive Success and Offspring Viability and Performance." *Animal Behaviour* 59, no. 2 (2000): 371–78.

Scelza, Brooke. "Choosy but Not Chaste: Multiple Mating in Human Females." *Evolutionary Anthropology: Issues, News, and Reviews* 22, no. 5 (2013): 259–69.

Walker, Robert S., Mark V. Flinn, and Kim R. Hill. "Evolutionary History of Partible Paternity in Lowland South America." *Proceedings of the National Academy of Sciences* 107, no. 45 (2010): 19195–200.

Brown, Gillian R., Kevin N. Laland, and Monique Borgerhoff Mulder. "Bateman's Principles and Human Sex Roles." *Trends in Ecology and Evolution* 24, no. 6 (2009): 297–304.

Baranowski, Andreas M., and Heiko Hecht. "Gender Differences and Similarities in Receptivity to Sexual Invitations: Effects of Location and Risk Perception." *Archives of Sexual Behavior* 44, no. 8 (2015): 2257–65.

Gowaty, Patricia Adair, Rebecca Steinichen, and Wyatt W. Anderson. "Mutual Interest Between the Sexes and Reproductive Success in *Drosophila pseudoobscura*." *Evolution* 56, no. 12 (2002): 2537–40.

Gowaty, Patricia Adair, Yong-Kyu Kim, and Wyatt W. Anderson. "No Evidence of Sexual Selection in a Repetition of Bateman's Classic Study of *Drosophila melanogaster*." *Proceedings of the National Academy of Sciences of the United States of America* 109, no. 29 (2012): 11740–45.

Trivers, Robert L. "Sexual Selection and Resource-Accruing Abilities in *Anolis garmani*." *Evolution* 30, no. 2 (1976): 253–69.

Janicke, Tim, et al. "Darwinian Sex Roles Confirmed Across the Animal Kingdom." *Science Advances* 2, no. 2 (2016).

Tang-Martínez, Zuleyma. "Rethinking Bateman's Principles: Challenging Persistent Myths of Sexually Reluctant Females and Promiscuous Males." *Journal of Sex Research, Annual Review of Sex Research Special Issue*. Published online April 13, 2016.

Chapter 7: Why Men Dominate

"Classification of Female Genital Mutilation." World Health Organization. http://www .who.int/reproductivehealth/topics/fgm/overview/en/.

"Prevalence of FGM." World Health Organization. http://www.who.int/reproductive health/topics/fgm/prevalence/en/.

Wardere, Hibo. *Cut: One Woman's Fight Against FGM in Britain Today*. London: Simon & Schuster, 2016.

Foreman, Amanda. "Why Footbinding Persisted in China for a Millennium." *Smithsonian Magazine*, February 2015. http://www.smithsonianmag.com/history/why -footbinding-persisted-china-millennium-180953971/?page=1.

Tapscott, Rebecca. "Understanding Breast 'Ironing': A Study of the Methods, Motivations, and Outcomes of Breast Flattening Practices in Cameroon." Feinstein International Center, May 2012.

Strassmann, Beverly I., et al. "Religion as a Means to Assure Paternity." *Proceedings of the National Academy of Sciences* 109, no. 25 (2012): 9781–85.

Hrdy, Sarah Blaffer. *The Woman That Never Evolved*. Cambridge, MA: Harvard University Press, 1981.

"Delhi Rapist Says Victim Shouldn't Have Fought Back." *BBC News*, March 3, 2015. http://www.bbc.co.uk/news/magazine-31698154.

Sherfey, Mary Jane. *The Nature and Evolution of Female Sexuality*. New York: Vintage Books, 1973.

Hrdy, Sarah Blaffer. "Raising Darwin's Consciousness: Female Sexuality and the Prehominid Origins of Patriarchy." *Human Nature* 8, no. 1 (1997): 1–49.

Troisi, Alfonso, and Monica Carosi. "Female Orgasm Rate Increases with Male Dominance in Japanese Macaques." *Animal Behaviour* 56, no. 5 (1998): 1261–66.

Pavlicev, Mihaela, and Günter Wagner. "The Evolutionary Origin of Female Orgasm." *Journal of Experimental Zoology* (2016).

Lerner, Gerda. *The Creation of Patriarchy*. New York: Oxford University Press, 1986.

Hrdy, Sarah Blaffer. *Mother Nature: Natural Selection and the Female of the Species*. London: Chatto & Windus, 1999.

Watkins, Trevor. "From Foragers to Complex Societies in Southwest Asia." In *The Human Past—World Prehistory and the Development of Human Societies*, edited by Chris Scarre, 201–33. London: Thames & Hudson, 2005.

Smuts, Barbara. "The Evolutionary Origins of Patriarchy." *Human Nature* 6, no. 1 (1995): 1–32.

Andics, Attila, et al. "Voice-Sensitive Regions in the Dog and Human Brain Are Revealed by Comparative fMRI." *Current Biology* 24, no. 5 (2014): 574–78.

Muller, Martin N., et al. "Male Coercion and the Costs of Promiscuous Mating for Female Chimpanzees." *Proceedings of the Royal Society B* 2074 (2007): 1009–14.

Stanford, Craig. "Despicable, Yes, but Not Inexplicable." Book review. *Scientific American*, November-December 2009. http://www.americanscientist.org/bookshelf/pub/despicable-yes-but-not-inexplicable.

———. "The Social Behaviour of Chimpanzees and Bonobos: Empirical Assumptions and Shifting Evidence." *Current Anthropology* 39, no. 4 (1998): 399–420.

Kemper, Steve. "Who's Laughing Now?" *Smithsonian Magazine*, May 2008. http://www.smithsonianmag.com/science-nature/whos-laughing-now-38529396/?no-ist.

De Waal, Frans B. M. "Bonobo Sex and Society." *Scientific American*, June 1, 2006. http://www.scientificamerican.com/article/bonobo-sex-and-society-2006-06/.

Parish, Amy R., and Frans B. M. De Waal. "The Other 'Closest Living Relative': How Bonobos (*Pan paniscus*) Challenge Traditional Assumptions about Females, Dominance, Intra- and Intersexual Interactions, and Hominid Evolution." *Annals of the New York Academy of Sciences* 907 (2000): 97–113.

White, F. J., and K. D. Wood. "Female Feeding Priority in Bonobos, *Pan paniscus*, and the Question of Female Dominance." *American Journal of Primatology* 69, no. 8 (2007): 837–50.

Ralls, Katherine. "Mammals in Which Females Are Larger Than Males." *Quarterly Review of Biology* 51, no. 2 (1976): 245–76.

Parish, Amy Randall. "Sex and Food Control in the 'Uncommon Chimpanzee': How Bonobo Females Overcome a Phylogenetic Legacy of Male Dominance." *Ethology and Sociobiology* 15, no. 3 (1994): 157–79.

Gowaty, Patricia Adair, ed. *Feminism and Evolutionary Biology: Boundaries, Intersections and Frontiers*. New York: Chapman & Hall, 1997.

Chapter 8: The Old Women Who Wouldn't Die
The inspiration for this chapter was an article on menopause published by the author in the *Observer* on March 30, 2014. Available online at http://www.theguardian.com /society/2014/mar/30/menopause-natures-way-older-women-sexually-attractive.
"Bethlem's Changing Population." *Bethlem Museum of the Mind* (blog). July 26, 2010. http://museumofthemind.org.uk/blog/post/life-in-a-victorian-asylum-2-clerks -and-governesses.
Odame-Asante, Emily. "'A Slave to Her Own Body': Views of Menstruation and the Menopause in Victorian England, 1820–1899." Dissertation, University College London, 2012.
Smith, R. Percy, Charles J. Macalister, and T. B. Grimsdale. "Discussion on the Psychoses of the Climacteric." *British Medical Journal* 2, no. 2707 (1912): 1378–86.
Rosenhek, Jackie. "Mad with Menopause." *Doctor's Review*, February 2014.
Ward, Suzie. "A History of the Treatment of the Menopause." Dissertation, Wellcome Institute for the History of Medicine at University College London, 1996.
Wilson, Robert A. *Feminine Forever*. London: W H Allen, 1966.
Santosa, Sylvia, and Michael D. Jensen. "Adipocyte Fatty Acid Storage Factors Enhance Subcutaneous Fat Storage in Postmenopausal Women." *Diabetes* 62, no. 3 (2013): 775–82.
Whitehead, Saffron. "Milestones in the History of HRT." *Endocrinologist* (Spring 2015): 20–21.
Fausto-Sterling, Anne. *Myths of Gender: Biological Theories About Men and Women*. 2nd rev. ed. New York: Basic Books, 1992.
Cooper, Wendy. *No Change: Biological Revolution for Women*. London: Hutchinson, 1975.
Bell, Susan E. "Changing Ideas: The Medicalization of Menopause." *Social Science and Medicine* 24, no. 6 (1987): 535–42.
Stone, Bronte A., et al. "Age Thresholds for Changes in Semen Parameters in Men." *Fertility and Sterility* 100, no. 4 (2013): 952–58.
Bosch, Mercè, et al. "Linear Increase of Structural and Numerical Chromosome 9 Abnormalities in Human Sperm Regarding Age." *European Journal of Human Genetics* 11 (2003): 754–59.
Williams, George C. "Pleiotropy, Natural Selection, and the Evolution of Senescence." *Evolution* 11, no. 4 (1957): 398–411.
Loudon, Irvine. "Maternal Mortality in the Past and Its Relevance to Developing Countries Today. *American Journal of Clinical Nutrition* 72, no. 1 (2000): 241–46.
Hawkes, Kristen, James F. O'Connell, and Nicolas G. Blurton Jones. "Hardworking Hadza Grandmothers." In *Comparative Socioecology: The Behavioural Ecology of Humans and Other Mammals*, edited by V. Standen and R. A. Foley, 341–66. London: Basil Blackwell, 1989.
Hawkes, K., J. F. O'Connell, N. G. Blurton Jones, H. Alvarez, and E. L. Charnov. "Grandmothering, Menopause, and the Evolution of Human Life Histories." *Proceedings of the National Academy of Sciences USA* 95 (1998): 1336–39.

Hawkes, Kristen, and James E. Coxworth. "Grandmothers and the Evolution of Human Longevity: A Review of Findings and Future." *Evolutionary Anthropology* 22 (2013): 294–302.

Thomas, Elizabeth Marshall. "Reflections: The Old Way." *New Yorker*, October 15, 1990. http://www.newyorker.com/magazine/1990/10/15/the-old-way.

"Life Expectancy." The King's Fund. http://www.kingsfund.org.uk/time-to-think-differently/trends/demography/life-expectancy. Accessed June 1, 2016.

"Mortality in the United States, 2014." US National Center for Health Statistics. http://www.cdc.gov/nchs/data/databriefs/db229.htm. Accessed June 1, 2016.

Hrdy, Sarah Blaffer. *Mothers and Others: The Evolutionary Origins of Mutual Understanding*. Cambridge, MA: Belknap Press of Harvard University Press, 2009.

Sear, Rebecca, and David A. Coall. "How Much Does Family Matter? Cooperative Breeding and the Demographic Transition." *Population and Development Review* 37 (2011): 81–112.

Shanley, D. P., R. Sear, R. Mace, and T. B. L. Kirkwood. "Testing Evolutionary Theories of Menopause." *Proceedings of the Royal Society B: Biological Sciences* 274, no. 1628 (2007): 2943–49.

Foster, Emma A., et al. "Adaptive Prolonged Postreproductive Life Span in Killer Whales." *Science* 337, no. 6100 (2012): 1313.

Brent, Lauren J. N., et al. "Ecological Knowledge, Leadership, and the Evolution of Menopause in Killer Whales." *Current Biology* 25, no. 6 (2015): 746–50.

Kuhle, Barry X. "An Evolutionary Perspective on the Origin and Ontogeny of Menopause." *Maturitas* 57, no. 4 (2007): 329–37.

Lahdenperä, Mirkka. "Severe Intergenerational Reproductive Conflict and the Evolution of Menopause." *Ecology Letters* 15, no. 11 (2012): 1283–90.

Lahdenperä, M., V. Lummaa, and A. F. Russell. "Menopause: Why Does Fertility End Before Life?" *Climacteric* 7, no. 4 (2004): 327–32.

Kachel, A. Friederike, L. S. Premo, and Jean-Jacques Hublin. "Grandmothering and Natural Selection." *Proceedings: Biological Sciences* 278, no. 1704 (2011): 384–91.

O'Connell, James F., Kristen Hawkes, and Nicolas B. Blurton Jones. "Grandmothering and the Evolution of *Homo erectus*." *Journal of Human Evolution* 36 (1999): 461–85.

Kim, Peter S., James E. Coxworth, and Kristen Hawkes. "Increased Longevity Evolves from Grandmothering." *Proceedings of the Royal Society B* (October 24, 2012), DOI: 10.1098/rspb.2012.1751.

Kim, P. S., J. S. McQueen, J. E. Coxworth, and K. Hawkes. "Grandmothering Drives the Evolution of Longevity in a Probabilistic Model." *Journal of Theoretical Biology* 353 (2014): 84–94.

Morton, R. A., J. R. Stone, and R. S. Singh. "Mate Choice and the Origin of Menopause." *PLOS Computational Biology* 9, no. 6 (2013).

Clancy, Kate. "Ladybusiness Anthropologist Throws Up Hands, Concedes Men Are the Reason for Everything Interesting in Human Evolution." *Scientific American* (blog). June 29, 2013. http://blogs.scientificamerican.com/context-and-variation/men-menopause-evolution.

Marlowe, Frank. "The Patriarch Hypothesis: An Alternative Explanation of Menopause." *Human Nature* 11, no. 1 (2000): 27–42.

Gurven, M., and H. S. Kaplan. "Beyond the Grandmother Hypothesis: Evolutionary Models of Human Longevity." In *The Cultural Context of Aging: Worldwide Perspectives*. 3rd ed. Edited by J. Sokolovsky, 53–66. Westport, CT: Praeger, 2009.

Tuljapurkar, S. D., C. O. Puleston, and M. D. Gurven. "Why Men Matter: Mating Patterns Drive Evolution of Human Lifespan." *PLOS ONE*, no. 8 (2007).

Tre, Lisa. "Men Shed Light on the Mystery of Human Longevity, Study Finds." Stanford News Service, September 12, 2007. http://news.stanford.edu/pr/2007/pr-men -091207.html.

Ayers, Beverley N., et al. "The Menopause." *Psychologist* 24 (2011): 348–53.

Im, Eun-Ok, Seung Hee Lee, and Wonshik Chee. "Sub-Ethnic Differences in the Menopausal Symptom Experience: Asian American Midlife Women." *Journal of Transcultural Nursing* 21, no. 2 (2010): 123–33.

Afterword

Monatgu, Ashley. *The Natural Superiority of Women*. New York: Macmillan, 1952.

Konner, Melvin. *Women After All: Sex, Evolution, and the End of Male Supremacy*. New York: W. W. Norton, 2015.

"Last Five Years Account for More Than One-Quarter of All Abortion Restrictions Enacted Since Roe." Guttmacher Institute, January 13, 2016. https://www.guttmacher .org/article/2016/01/last-five-years-account-more-one-quarter-all-abortion -restrictions-enacted-roe.

Rúdólfsdóttir, Annadís Greta. "Iceland Is Great for Women, but It's No Feminist Paradise." *Guardian*, October 28, 2014. http://www.theguardian.com/commentisfree /2014/oct/28/iceland-women-feminist-paradise-gender-gap-pay.

Gracia, Enrique, and Juan Merlo. "Intimate Partner Violence Against Women and the Nordic Paradox." *Social Science and Medicine* (May 2016).

Tavris, Carol. *The Mismeasure of Woman: Why Women Are Not the Better Sex, the Inferior Sex, or the Opposite Sex*. New York: Simon & Schuster, 1992.

INDEX

abortion: of female fetuses, 29–31; restricting access to, 105, 178
Aché nomadic hunter-gatherers, Paraguay: gender equality, 117; hunting by females, 119; hunting by males, 111–12; role of grandmothers, 164
The Age of Scientific Sexism (Ruti), 9
alleles (genes), 39
alloparents: and cooperative breeding, 105; defined, 102; grandmothers as, 165–66
American Anthropological Association, male bias, 108–109
amygdala, sex differences in, 86, 92
Anderson, Wyatt, 134–35
androgen, 25–26, 59–61. *See also* sex hormones
"The Angel in the House" (Patmore), 16
animals, research involving: animal orgasm studies, 145; applicability to humans, 58; chromosome research on mice, 41; impacts of sex hormones on brain growth and behavior, 55–57; and male guarding behaviors, 143; sex

biases in, 43; species showing female dominance, 151–53. *See also* primatology
anthropology: contributions to understanding of gender identity, 26–27; focus on male behaviors, 107–10; and the hunting hypothesis, 107–13, 115; on male contributions to human longevity, 172; and studies of modern-day hunter-gatherer societies, 101–2, 117; and women as tool inventors and users, 110
archaeology, data on sex differences from, 94–95
Ardrey, Robert, 108
Arnold, Arthur, 38–41, 43–44, 46
Ashworth, Ann, 31
Asia, South Asia: cultural preferences for male children, 29–32, 104, 178; extended families in, 163
Austad, Steven, 33–37, 40, 42–43
autism: as extreme version of the systemizing brain, 52, 54–55; relationship with fetal sex hormones, 68–69

autoimmune disease, sex differences in, 36–37
Ayers, Beverley, 174

baboons, 99, 154
Baranowski, Andreas, 132
Baron-Cohen, Simon: empathizing-systemizing theory, 52–55, 63, 67–68; studies of fetal testosterone and brain development, 51–52, 66–69; study on gender differences in newborns, 53–54
Bateman, Angus John: critiques of fruit fly study, 127–28, 132, 134–35; fruit fly mating studies, 121–25, 136–37; and sexual selection theory, 129
"beauty map" of British women (Galton), 17
Beery, Annaliese, 43
Behan, Peter, 56–57
behavioral research, 49–54, 129–31
Bennett, Craig, on functional magnetic resonance imaging, 82–83
Berthold, Adolph, experiments with cockerel testes, 22–23
Bethlehem Royal Hospital (Bedlam), postmenopausal women in, 157–58
biological research, 14–15, 36, 42–43
Bird, Rebecca Bliege, 112, 118–19
Blackwell, Antoinette Brown, 18
Blair-Bell, William, 25
bluebirds, 134
Bluhm, Cynthia, 130
bonobos: bonds between females, 100, 153; casual sex among, 152–53; dominance of females over males, 150–52; fertility patterns and behaviors, 102–3; hunting by females,

153; identification as separate species, 151
"Boys Will Be Boys" (Pinker), 125–26
brains, human: composition and architecture, 77–78; diversity and uniqueness, 90–92; neural connections in, 78–79; plasticity, 89–92; size, as ratio to body size, 76
brains, sex differences research: blood flow studies, 76–77; critiques of, 77–78, 81–82, 85–86; dimorphism assumptions, 84; and gender stereotyping, 88, 90, 93; and hippocampus size, 84–85; weight and volume studies, 72–76. *See also* Baron-Cohen, Simon
breast ironing, 142
Bribiescas, Richard Gutierrez, 102–3, 105–6, 111
Brown, Gillian, 131
Brown-Séquard, Charles-Édouard, 23
Burnell, Jocelyn Bell, 9
Buss, David, 125

Cahill, Larry, 86–87, 89
Cameroon, West Africa, breast ironing in, 142
Carosi, Monica, 145
Cerebral Dominance (Geschwind and Galaburda), 57
chess, dominance of males in, 87
childbirth: and cultural restrictions on female equality, 119; differences between primates and humans, 102; and female failure to evolve, 15; mortality associated with, 163
children: early environment and brain development, 112–13; and emergence of gender identity,

50–51; external stimulation and brain function, 71–72; factors influencing survival, 166; importance of community supports, 106; male, cultural preferences for, 29–30, 32, 104, 178; parental investment in, 123–24; role of parents in gender socializing, 63, 71; sex differences in health and physiology, 31–33; survival of, and grandmother effect, 163, 166. *See also* infanticide, feticide

chimpanzees: birthing behaviors, 102; dominance of males among, 97; female tool-using skills, 110; male coercion of females by, 149–50; male preference for older females, 170

China, foot binding in, 142

"Choosy But Not Chaste: Multiple Mating in Human Females" (Scelza), 130–31

Cimpian, Andrei, 66

Clark, Russell, 120–21, 127, 132–33

Clayton, Janine, 44–45, 47

clitoris removal, during female genital mutilation, 139–40

Coall, David, 106, 166

Coates, John, 27

cockerel testes, 22–23

cognitive neuroscience, 82. *See also* neuroscience

Colom, Roberto, 65

complementarity principle, 17, 80–81, 94

Confucius, 142

congenital adrenal hyperplasia, 63

Connellan, Jennifer, 53–54, 66–68

consciousness-raising, 134

Cooper, Wendy, 160

cooperative breeding systems, 102–5, 107, 115–16

Craig, Michael, 104

The Creation of Patriarchy (Lerner), 146–47

Crittenden, Alyssa, 171–73

Croft, Darren, 166–67

Cronin, Helena, 51–52

cultural/social factors: and the encouragement of high-achieving males, 65–66; and excess mortality among girl babies, 32; and female vs. male response to disease, 37–38; impact on female equality, 119

Curie, Marie, 8

Cut: One Woman's Fight Against FGM in Britain Today (Wardere), 141

Darwin, Charles: assumptions about male superiority, 14, 18, 95, 107; Kennard's letter to, 13–16; and sexual selection theory, 122

Datoga pastoralist-warriors, 106

Day, Alice Chenoweth. *See* Gardener, Helen Hamilton

de Beauvoir, Simone, 13

Delusions of Gender (Fine), 67, 84

The Descent of Man, and Selection in Relation to Sex (Darwin), 14–15, 121

de Waal, Frans, 153

diffusion tensor imaging, 78

"digging sticks," 110

digoxin, sex-related research findings, 45–46

Disteche, Christine, 40

Do Babies Matter: Gender and Family in the Ivory Tower (Mason, Wolfinger, and Goulden), 4

Dogon communities, Mali, 142

"Do Men Need to Cheat on Their
    Women? A New Science Says Yes"
    (*Playboy*), 124
ducks, mallard, 130
Dyble, Mark, 116–17

Eddy, Sarah, 5–6
education, sexism in, 8
Ehrenberg, Israel, 177
Eliot, Lise, 85
empathizing-systemizing theory
    (Baron-Cohen): critiques, 63,
    67–70; supporting evidence,
    popularity, 52–55, 57, 63
endocrinology. *See* estrogen; hormone
    therapy for menopause; sex hor-
    mones; testosterone
Engels, Friedrich, 146
the Enlightenment, view of science
    during, 16
Equal Pay Act, UK, and the gender pay
    gap, 5
The Essential Difference (Baron-Cohen),
    54–55
essentialism, 93
Estioko-Griffin, Agnes, 114–15
estrogen: loss of, and menopausal
    symptoms, 159–60; in men, discov-
    ery of and implications, 25–26. *See
    also* hormone replacement therapy
    for menopause; menopause; sex
    hormones
Eté, Democratic Republic of the
    Congo, alloparenting among, 102
Evans, Herbert, 26
Eve, as subservient woman, 19
evolutionary biology: data on sex dif-
    ferences, 94–95; and the develop-
    ment of language and intelligence,
    112–13; explanations for female

orgasm, 145; explanations for
    menopause and postmenopausal
    survival, 161–63, 165, 168–69;
    and the importance of primate re-
    search, 98–99, 154; sexist assump-
    tions, 19, 14–22, 98–99, 116–17,
    134, 136; and sexual selection
    theory, 121–25
evolutionary psychology, and gender-
    based concepts of monogamy and
    polygamy, 125–26
*The Evolution of Desire: Strategies of Hu-
    man Mating* (Buss), 126
The Evolution of Human Sexuality (Sy-
    mons), 125
The Evolution of Sex (Geddes and
    Thomson), 17
*The Evolution of Woman, an Inquiry into
    the Dogma of Her Inferiority to Man*
    (Gamble), 20
extended families, and the grandmother
    hypothesis, 163
extended longevity hypothesis, 165, 168

Facts and Fictions of Life (Gardener), 74
fathers, fathering, 103, 106–7. *See also*
    alloparents; partible patrimony
Fausto-Sterling, Anne: on fetal sex hor-
    mones and brain development, 70;
    on human beings as developmental
    systems, 70; newborn and baby
    research, 55, 71–72; on Victorian
    concepts of femininity, 25; on
    Wilson's sexist language, 160
female dominance, animals that show,
    151–53
female genital mutilation (FGM),
    139–41
females, women: and allopar-
    ents, 101–2; biases against in

high-achieving disciplines, 2–5, 66; as biologically predetermined, 3, 120–21, 131, 133, 143; child-care role, and development of language, 112–13; and choice of mate, benefits to children, 130; and concepts of femaleness, femininity, 16, 23–28, 90; cooperation among, 156; disease incidence and virulence in, 36–37, 40–41; economic limitations and restrictions, 17–18; educational limitations and restrictions, 8; endurance and strength, 31–33, 113–14, 177; experience of, brain effects, 89; as gatherers, work involved in, 109–10; and the gender pay gap, 5; as hunters, 110, 114–15; intelligence and skill acquisition, 63–65, 72–76, 84, 90, 110; and the maternal instinct, 103–4; and mate selectivity, 133; monthly cycles, physiology of, 159; as natural leaders, 177–78; as naturally monogamous, 121–26; pro-male gender bias shown by, 5; sexual assertiveness, 128; sex-related response to medications, 44–45; unique characteristics, 61–62; unpaid labor performed by, 4–5; violence against, 178–79. *See also* empathizing-systemizing theory; menopause; sex hormones; sexuality, female; virginity, female chastity

Feminine Forever (Wilson), 159–60

feminism, contributions to the practice of good science, 3, 10–12, 21–22, 27, 74, 128, 134, 179–81

Feminist Approaches to Science (Hrdy), 128

fetal brain development, impact of sex hormones, 55–58, 68–70

fetuses, as initially female, 23–24

FGM. *See* female genital mutilation (FGM)

Fields medal for mathematics, 2

Fine, Cordelia, 67, 84, 89–90

five-alpha-reductase deficiency, 59

Flinn, Mark, 107, 131

Flint, Marcha, 174

Florida State University experiment on casual sex, 120, 132–33

follicles: depletion of, 168; monthly release of, 159

foot binding, China, 142

Foreman, Amanda, 142

Fossey, Dian, 96

Franklin, Rosalind, 9

From Eve to Evolution: Darwin, Science, and Women's Rights in Gilded Age America (Hamlin), 19

functional magnetic resonance imaging, 81–83

Galaburda, Albert, 56–57

Galton, Francis, 17

Gamble, Eliza Burt, 19–22, 27, 74, 108

Gardener, Helen Hamilton (Alice Chenoweth Day), 74–76, 81

gathering activities, 109–10

Geddes, Patrick, 17

Geertz, Clifford, 125

gender: and limitations on women's work, 16; women's rights movement, 16–17, 21

gender, gender identity: as biologically determined, 3, 12, 18, 27, 48, 52–54, 80–81; complementarity principle, 17, 80–81, 94; emergence of, in early childhood,

50–51; gender similarities hypothesis, 61, 64; in human embryos, 24; and neurosexism, 84; in newborns, 53–54; and primate studies, 55–56; and sexual behaviors, 120–21; sex vs., 28; as socially and culturally determined, 14–17, 19–20, 28, 50–51, 56, 88–93, 127, 146, 180; spectrum for, 26, 70. *See also* females, women; males, men; sex hormones; sexuality, female

"Gender Differences in Receptivity to Sexual Offers" (Clark and Hatfield), 121

genetic research, 38–41

genitals, sex effects, 92

Geschwind, Norman (Geschwind-Behan-Balabura theory), 56–57

gibbons, 155

Gilman, Charlotte Perkins, 18, 74

Gliga, Teodora: on challenges of baby research, 50–51; on gender differences as a continuum, 70; on the importance of replication, 67–68; on sex differences in performance, 64

Goodall, Jane, 96, 110, 149–50

Goulden, Marc, 4

Gowaty, Patricia, 130, 133–37

Goy, Robert, 55–56

the grandmother hypothesis: critiques of, 173; Gurven's two-sex model for, 172; Hawkes's research, 163–65, 168–69; survival benefits, 166, 167; Williams' elaboration of, 163

Griffin, Bion, 114–15, 118

Grossi, Giordana, 66–67

Grunspan, Dan, 5–6

Guillebaud, John, 38

Gur, Raquel, 78–80, 84, 88

Gur, Ruben: complementarity hypothesis, 79–81; critiques of, 84, 88; on sex differences in the human brain, 76–79, 93; support for, 85–86

Gurven, Michael: on bias in human research, 173; on male contributions to human longevity, 172; on male nipples, 170; revision of hunting hypothesis, 111–12, 118

Hadza hunter-gatherers, Tanzania: alloparents among, 102; differing anthropological perspectives on, 164; fathering among, 106; gender equality, 117; male hunting, 111; role of older women, 171

Halpern, Diane, 90–91

Hamadryas baboons, 149

Hamlin, Kimberly, 7, 19, 21

Hammond, William Alexander, 74–76

Hanuman langur monkeys, 97–98, 149, 154

"Hardworking Hadza Grandmothers" (Hawkes), 164

Hatfield, Elaine, 120–21, 127, 132–33

Hawkes, Kristen: critiques of, 112, 173; on feminist contributions to science, 12; and the grandmother hypothesis, 163–64, 168–69, 174–75; on hunting as a reliable food source, 111; support for, 171, 174

health differences, female vs. male, 32, 34–41, 45–47

Heape, Walter, 21–22

Hecht, Heiko, 132–33

Hill, Kim, 107, 111–12, 117–18, 131

Himba nomadic farmers, Namibia, 129–30, 132

Hines, Melissa: on replication in science, 62; research with intersex

people, 61–62; studies of impacts
of prenatal testosterone, 58, 69, 72;
studies of sex differences in intelligence and behavior, 63–65, 84
hippocampus, sex-related size differences, 84–85
Holmes, Donna, 173
hormone replacement therapy for
menopause, 160–61
hormones, role of, 23. *See also* sex
hormones
Hrdy, Sarah Blaffer: challenges to
gender stereotypes, 98–99, 103–4,
142–43, 147–48; cooperative
breeding systems, 102; on language
development in humans, 112; on
multiple mating, 127; primatology research and findings, 96–98,
100–101, 103–4; resistance to
sexism in science, 99–100, 128; and
the role of grandmothers, 165–66
human beings, as developmental systems, 70–71
hunter-gatherer societies: alloparents,
102; anthropological studies of,
101–2; and the division of workloads, 113–14; egalitarian, 116–18;
and female hunters, 114–15,
118–19; importance to human
evolutionary history, 107–8
hunting: by females, evidence from
chimpanzees, 110–11; by females
among the Nanadukan Agta,
114–15; by males, assumptions
about, 3, 107–13; and strength required of female gatherers, 113; as
variable food source, implications,
111; view of as incompatible with
motherhood, 118
The Hunting Hypothesis (Ardrey), 108

Hurtado, Ana Magdalena, 119
Hyde, Janet Shibley, 64

immune system, female, 35–36
India, preference for male children,
29–30
infanticide, feticide, 97–98, 104, 178
infants. *See* children
infibulation, 140
intelligence, sex-related differences,
63–65, 72–76, 84, 90, 110, 112–13
intersex people, 59–61
intuitiveness, as female trait, 80. *See also*
empathizing-systemizing theory

Jacklin, Carol Nagy, 64
Jacobson, Anne Jaap, 92–93
Japanese macaques, 145
Joel, Daphna, 84, 91–93
Jordan-Young, Rebecca, 89–90

Kachel, Friederike, 168
Kaiser, Anelis, 89–90
Kennard, Caroline, 13–16
Khurana, Mitu, 29–30
Kidd, Celeste, 112–13
killer whales (orcas), 162, 166–67
Kim, Yong-Kyu, 135
Klein, Sabra, 37
Konner, Melvin, 117–18, 146, 177–78
Kuhle, Barry, 167–68
!Kung hunter-gatherers, southern
Africa: alloparents among, 102;
hunting by males, 111; role of
women, 109, 113

Laland, Kevin, 131
Lancaster, Chet, 108
language development in humans, 112
Lawn, Joy, 31–32, 34

Lee, Richard, 109, 113
left-handedness, 57
Lerner, Gerda, 146–47
Leslie, Sarah-Jane, 66
life expectancy, sex differences in, 33, 165, 168–72
life-span-artifact hypothesis, 165
Linton, Sally (Sally Slocum), 108–10, 112
Lummaa, Virpi, 167–68, 170
Lutchmaya, Svetlana, 68
Lyon, Mary Frances, 40

Maccoby, Eleanor, 64
Maguire, Eleanor, 89
males, men: behavioral differences, baby research, 50; brain size studies, 72–73; disease incidence and virulence, 34, 36–37, 39–41; hormonal response to contact with babies, 105; intelligence of, comparisons with females, 64–65, 75; mate guarding behaviors, 141–43; as naturally polygamous, 121–26; and the patriarch theory, 167, 170–71; preference for younger women, 169–71, 173; response to medications, 44–46; sexual behaviors, 120–21; and sexual insecurity, 137–38, 146, 178; sperm activity as metaphor for, 17; stereotypes associated with, 5, 9, 16, 23–26, 52, 61–64, 66, 84, 110; and the Y chromosome, 39
malnutrition, 38
Man the Hunter (Washburn and Lancaster), 108
"Man the Hunter" symposium, University of Chicago, 107–8
Marlowe, Frank, 170–73

Martin, Carol Lynn, 50–51
Martu hunter-gatherers, Western Australia, 119
Mason, Mary Ann, 4
mate guarding behaviors. See sexual jealousy, mate guarding
maternal instinct, myths about, 103–4
math, math achievement: changing sex ratios associated with, 90; underrepresentation of women, 2, 65–66
Matthews, Paul: on cognitive neuroscience, 82; on diffusion tensor imaging, 78; on early research errors, 83–84; on inter-individual variability in brain function, 93; on studies of brain plasticity in adults, 89; study of toy type and brain development, 76, 90
Mbendjele hunter-gathers, Democratic Republic of the Congo, 116–17
McEwen, Bruce, 55–56
McManus, Chris, 57
Mead, Margaret, 26–27
medical research: binary nature of, 47; exclusion of women from, rationale, 43; focus on males, 43; implications for women's health and treatment, 44–45, 48; requirement to include women as test subjects, 47; sex differences in treatment response, 44–47
Meitner, Lise, 8–9
men. See males, men
Men: Evolutionary and Life History (Bribiescas), 105–6
menopause: alternatives to the grandmother hypothesis, 168; changing attitudes towards, 161, 174; and depletion of sex hormones, 159; evolutionary origins, 165, 169–73;

evolution of, factors contributing to, 162–67; first recorded mention, 165; historical misunderstanding and fear of, 158; and hormone replacement therapy, 160–61; medicalization of, 160, 174; in nonhuman species, 162; as protective, 163

menstruation, 35–36, 38

Meriam hunter-gatherers, Torres Strait Islands, 118

Mesopotamia, ancient, female subjugation in, 146–47

Miller, Amber, 113

Miller, David, 90–91

Miller, Geoffrey, 126

The Mind Has No Sex? Women in the Origins of Modern Science (Schiebinger), 7

The Mismeasure of Woman: Why Women Are Not the Better Sex, the Inferior Sex, or the Opposite Sex (Tavris), 179–80

modesty, female. *See* virginity, female chastity

monogamy: evolution of, factors contributing to, 125–26; and partible patrimony, 107; and sexual selection theory, 121–25; stereotypes associated with, 121–22

Montagu, Ashley, 29, 176–78

Montagu, Mary Wortley, 177

Morton, Richard, 169–171, 173

Moss-Racusin, Corinne, 5

Mosuo society, China, 130–31

motherhood, mothers: and alloparents, 101–2, 165–66; hormonal response to contact with babies, 105; humans compared with primates, 100–101; primate mothers,

101; role in gender-socializing of infants, 71–72; stereotypes about, 118. *See also* children; the grandmother hypothesis

Mother Nature (Hrdy), 146–48

Mothers and Others: The Evolutionary Origins of Mutual Understanding (Hrdy), 101–2, 165–66

Mulder, Monique Borgerhoff, 131

Muller, Martin, 106, 149

multiple mating among females, 127, 129–31

Nanadukan Agta hunter-gathers, Luzon, Philippines, 114–16, 118

Nash, Alison, 66–67

National Institutes of Health Revitalization Act, 47

The Natural Superiority of Women (Ashley Montagu), 176, 178

The Nature and Evolution of Female Sexuality (Sherfey), 144–46

Nepal, work expected of female children in, 30

neurofeminism, 92–93

neuroscience: neurosexism, 83–84; and sex differences research, 75, 77, 83–84; technological innovations, implications, 81–82

Nobel Prize, 2

No Change: Biological Revolution for Women (Cooper), 160

Noether, Emmy, 8

the Nordic Paradox, 179

nuclear families, 107

O'Connell, James, 111

O'Connor, Cliodhna, 88

Oertelt-Prigione, Sabine, 35–38, 40–42, 47

orangutans, male dominance, 149
orgasm, female, 125, 144–45
Oudshoorn, Nelly, 24, 26
the ovum, female egg, 17
oxytocin, and response to babies, 105

Palanan Agta hunter-gatherers, Philip-
    pines, 116–17
Paleofantasy (Zuk), 113
"Parental Investment and Sexual Selec-
    tion" (Trivers), 123
Parish, Amy, 150–54, 156
partible patrimony, 107, 131
Patmore, Coventry, 16
patriarchal societies, establishment of,
    146–49
patriarch theory of male longevity
    (patriarch hypothesis), 170–72
Pavlicev, Mihaela, 145
pharmacology research, focus on males,
    43
physics, women working in, 2
Piaget, Jean, 50
Piantadosi, Steven, 112–13
pigeons, male dominance in, 137–38
Pinker, Steven, 51–52, 125–26
plasticity, brain, 89–92
PLOS ONE, apology for gender bias in
    editorial policy, 6
polygamy, 121–26
postmenopausal women: healthy
    and active, challenges posed by,
    161–62, 165–66, 173; and hormone
    treatments, 159; incarceration,
    157–58; infertility of, 158–59. *See
    also* the grandmother hypothesis;
    menopause patriarch theory of
    male longevity
"preferential-looking" experiments, 53
pregnancy, 32–33, 35–36, 113

primates: close relationship of mothers
    with children, 100–101; fertility,
    102; gender-cooperative species,
    155; life expectancy among, 165;
    males, dominance behaviors, 148–49
primatology: female researchers,
    96; focus on chimpanzees, 151;
    importance to human evolutionary
    history, 94–95, 154; male focused
    research, 97–100; and the origins
    of patriarchy, 148–49
property, women as, and sexual repres-
    sion of women, 146–48
psychologists, evolutionary, data on
    sex-related differences, 94–95
The Psychology of Sex Differences (Mac-
    coby and Jacklin), 64
Puleston, Cedric, 172

Quinton, Richard, 23–24, 27, 59–60

Radcliffe, Paula, 113
Rademaker, Marius, 44
Ralls, Katherine, 155
replication of scientific research, im-
    portance, 62, 67–68, 135
reproductive cost hypothesis for meno-
    pause, 168
rhesus macaque, 149
Richardson, Sarah, 42, 45, 47
right-brain development, 56–57
Rippon, Gina: on brain plasticity stud-
    ies, 89–90; on complementarity,
    81, 87; on neurosexism, 83–84;
    on the political implications of
    research, 87–88; on uniqueness of
    each brain, 91
Romanes, George John, 17, 75–76
Rosenberg, Karen, 102
Rousseau, Jean-Jacques, 80–81

Royal Society of London, election of
women to, 7
Ruble, Diane, 50
running, endurance, 113
Ruti, Mari, 9

Sandberg, Kathryn, 34–36, 38, 45
Saudi Arabia, women in, 143–44
Scelza, Brooke, 128–33
Schiebinger, Londa, 7–8, 16
scientific research: blind experimental
designs, 66; as more gray than
black and white, 11–12; myth of
objectivity of, 55, 87; replication
and, 62, 67–68, 135; on sex differences, 9–10, 56–58, 62; sexism in
research approaches and findings,
3–4, 9, 21, 56, 87, 99; standard
deviation and significance, 62–63;
by women, 2, 4–5, 7–9, 65–66. *See
also specific scientific disciplines*
Sear, Rebecca, 106, 166, 170
senescence hypothesis for menopause,
168
sex, gender vs., 28
Sex Antagonism (Heape), 21–22
sex chromosomes, 38–42
sex hormones: behavioral impacts,
55–56; behavioral impacts, animal
research, 55–56; biological function, 23–24; and brain development, 52, 55–58, 68–70; and
definitions of masculinity and
femininity, 25; discovery of, 23, 26;
female, and immune system advan-
tages, 35–37; and gender identity,
25–26, 61; intersex individuals,
59–61; ongoing research related to,
27; as treatments, 24–25, 159. *See
also* menopause

"Sex in Brain" talk (Gardener), 74
*Sex Itself: The Search for Male and Female
in the Human Genome* (Richardson),
41–42, 45
sex ratios, skewing of towards males, 30
Sexual Differentiation of the Brain (Goy
and McEwen), 56
sexual dimorphism, 93
sexuality, female: and assumptions about
norms, 121; and breast ironing, 142;
Engels's views on, 146; and female
genital mutilation, 139–41; and
female orgasm, 125, 144–45; and
foot binding, 142; forced marriage,
domestic violence and rape, 143–44;
and Hrdy's views on, 144–45; and
infidelity, Scelza's findings, 128–29;
and menstrual huts, 142; and moral
double standards, 143–44, 146–48;
and Sherfey's views on, 144,
145–46; and slavery, 147; stereotypes associated with, 120–21, 128;
and women as property, 146–48. *See
also* sexual jealousy, mate guarding;
virginity, female chastity
sexual jealousy, mate guarding, 138,
141, 143, 146–47, 149
sexual selection theory, 121–25, 128,
135–37
Seymour, Jane Katherine, 25
Sherfey, Mary Jane, 144–46
Short, Nigel, 87
Singh, Rama, 169–71, 173
Smuts, Barbara, 148–51, 154
Somalia, female genital mutilation in,
139
South Asia. *See* Asia, South Asia
spatial processing: as male skill, 80; role
of white matter in, 79
Spitzka, Edward, 75

standard deviation, 62–63
Stanford, Craig, 150, 153
Starin, Dawn, 101, 103, 127
Steinem, Gloria, 157
Steinichen, Rebecca, 134–35
Stevens, Nettie Maria, 8
Stone, Jonathon, 169–71, 173
subsistence living, strength and endurance required for, 113
Summers, Lawrence, 2, 51–52, 65
survival, females vs. males, 32–37, 41, 114
Symons, Don: critique of Gowaty's research, 135–36; critique of Hrdy's research, 128; critique of Sherfey's research, 144; sexual selection theory, 125–26

Tang-Martínez, Zuleyma, 127–28, 136–37
Tapscott, Rebecca, 142
Tavris, Carol, 179–80
testosterone, 25, 27. See also sex hormones
Thomas, Elizabeth Marshall, 164–65
Thomson, John Arthur, 17, 22
titi and tamarin monkeys, 103, 155
tools, female vs. male use of, 110
toys: sex preferences, 3, 68, 71; sex-typed toys, 62–63; types of, impact on brain development, 76, 90
Trevathan, Wenda, 102
Trivers, Robert: critique of Gowaty's research, 135–36; critique of Hrdy's research, 99; pigeon observations, 137–38; on sexual desire and reproductive capacity, 129–30; and sexual selection theory, 123–25
Troisi, Alfonso, 145
Tuljapurkar, Shripad, 172

Verma, Ragini, 80
A Vindication of the Rights of Woman (Wollstonecraft), 18
violence against women. See sexual jealousy, mate guarding; virginity, female chastity
virginity, female chastity: approaches to insuring, 141–42; emphasis on in patriarchal societies, 143–44, 146–48; importance of religious beliefs, 143–44; and violence against women, 178–79. See also sexuality, female
vote, right to, biological arguments against, 16–17, 21

Wagner, Günter, 145
Walker, Robert, 107, 131
Wardere, Hibo, 139–41
Washburn, Sherwood, 108
Whitehead, Saffron, 160–61
Why Can't a Woman Be More Like a Man? (Wolpert), 55
"Why Do Men Hunt?" (Gurven and Hill), 112
"Why Men Matter: Mating Patterns Drive Evolution of Human Life-span" (Gurven, Tuljapurkar, and Puleston), 172
wife, role of, 16–17, 25, 80, 87
Wilder, Burt Green, 76
Williams, George, 162–63
Wilson, Robert, 159–60
Wolf, Naomi, 120
Wolfe, Albert, 21–22
Wolfinger, Nicholas, 4
Wollstonecraft, Mary, 18, 49, 139
Wolpert, Lewis, 55
The Woman That Never Evolved (Hrdy), 128, 143

Woman the Gatherer (Dahlberg), Zihl-
man chapter in, 109, 113
"Woman the Gatherer: Male Bias in
Anthropology" (Linton), 108–9
women. *See* females, women
*Women After All: Sex, Evolution, and the
End of Male Supremacy* (Konner),
177–78
Wrangham, Richard, 154

X-linked disorders, 39–41, 65

Yalow, Rosalyn Sussman, 96
Yamanaka, Miki, 31

Zihlman, Adrienne, 109–11, 113–14
zolpidem (ambien), metabolism of, 46
Zondek, Bernhard, 26
Zuk, Marlene, 113